Climate Agendas and Instability

This book examines the expansion of climate governance frameworks in the EU and US and their re-framing as part of green industrial programs.

Addressing research on how vectors of stability and punctuation interact to advance or block policy progress, the book breaks new ground by presenting a theoretical framework suitable to integrate insights of comparative research and "sui generis" accounts of climate policy as a variable and multi-dimensional issue. In its empirical part, it compares two contrasting trajectories of climate policy-making: namely, the adaptation of the European Green Deal agenda to the exogenous shocks of the Covid pandemic and war in Ukraine through its NextGen and REPowerEU programs; and the launch of green industrial policies targeting infrastructure (Bipartisan Infrastructure Law) and inflation reduction (Inflation Reduction Act) in the US. Finally discussing to what degree the EU and US show signs of convergence towards a new type of climate policy from opposed starting points, the book identifies future research agendas around the topics of climate policy integration and politicization.

This book will be of key interest to scholars, students and practitioners of climate change governance, EU and US politics, environmental politics and comparative politics.

Frank Wendler is a Senior Researcher at the Center for Sustainable Society Research (CSS) and Senior Faculty Member (Privatdozent) at the Department for Economics and Social Sciences at the University of Hamburg, Germany.

Routledge Research in Comparative Politics

Coalition Politics in Central Eastern Europe
Governing in Times of Crisis
*Edited by Torbjörn Bergman, Gabriella Ilonszki
and Johan Hellström*

Collegial Democracy versus Personal Democracy
'We' the People or 'I' the People?
Edited by Chen Friedberg and Gideon Rahat

Pandemic-Era Civil Disorder in Post-Communist EU Member States
Joanna Rak

British and American Electoral Politics in the Age of Neoliberalism
Parallel Trajectories
Gerald Sussman

Constitutionalization of Politics in Comparative Perspective
Edited by Paweł Laidler, Dariusz Stolicki, Łukasz Jakubiak and Jacek Sokołowski

Climate Agendas and Instability
Green Recovery Programs and Policy Change in the EU and US
Frank Wendler

For more information about this series, please visit: https://www.routledge.com/
Routledge-Research-in-Comparative-Politics/book-series/CP

Climate Agendas and Instability

Green Recovery Programs and
Policy Change in the EU and US

Frank Wendler

Routledge
Taylor & Francis Group

LONDON AND NEW YORK

First published 2025
by Routledge
4 Park Square, Milton Park, Abingdon, Oxon OX14 4RN

and by Routledge
605 Third Avenue, New York, NY 10158

Routledge is an imprint of the Taylor & Francis Group, an informa business

British Library Cataloguing-in-Publication Data
A catalogue record for this book is available from the British Library

Library of Congress Cataloging-in-Publication Data
Names: Wendler, Frank, 1974– author.
Title: Climate agendas and instability : green recovery programs and policy change in the EU and US / Frank Wendler.
Description: Abingdon, Oxon ; New York, NY : Routledge, 2025. |
Series: Routledge research in comparative politics |
Includes bibliographical references and index.
Identifiers: LCCN 2024037054 (print) | LCCN 2024037055 (ebook) |
ISBN 9781032588995 (hardback) | ISBN 9781032589008 (paperback) |
ISBN 9781003452041 (ebook)
Subjects: LCSH: Green New Deal–European Union countries. |
Green New Deal–United States. | Climatic changes–Government policy–European Union countries. | Climatic changes–Government policy–United States. |
Industrial policy–Environmental aspects–European Union countries. |
Industrial policy–Environmental aspects–United States.
Classification: LCC GE190.E85 W38 2025 (print) |
LCC GE190.E85 (ebook) | DDC 363.7/0591094–dc23/eng/20240828
LC record available at https://lccn.loc.gov/2024037054
LC ebook record available at https://lccn.loc.gov/2024037055

ISBN: 978-1-032-58899-5 (hbk)
ISBN: 978-1-032-58900-8 (pbk)
ISBN: 978-1-003-45204-1 (ebk)

DOI: 10.4324/9781003452041

Typeset in Times New Roman
by Newgen Publishing UK

Contents

Illustrations

Figures

Tables

1 Introduction

New dynamics in climate politics and policy-making

The current governance and politics of climate change are transformed by two related dynamics. The first is an expansion of governance frameworks and agendas towards the definition of climate action as a cross-cutting rather than sectoral issue. A key element of this dynamic has been the proclamation of green deal agendas as a political concept to declare the target of carbon neutrality as a program for deep transformative change across society (Aronoff et al. 2019, Pettifor 2020, Ajl 2021, Rifkin 2020, Bloomfield & Stewart 2020, Samper et al. 2021, Mastini et al. 2022). As a result of commitments to the target of net-zero emissions of greenhouse gases (GHGs) across policy fields, climate change has left behind its status as a subfield of environmental policy and is now debated as a task of far-reaching change in virtually any field of policy-making. The concept of a green deal is associated particularly with climate policies in the EU through its European Green Deal (EGD) agenda (Skjaerseth 2021, Oberthür & von Homeyer 2023, Eckert 2021, Domorenok & Graziano 2023, Bongardt & Torres 2022, Delbeke & Vis 2019, Hainsch et al. 2022). However, similar developments are observed in North America, where the Biden administration has delivered a pledge to work towards carbon neutrality by mid-century and has declared the challenge of mitigating the climate crisis as a cross-government issue (Bang 2021, UNFCCC 2021). As a consequence, the political and institutional boundaries of climate governance and linkages between involved policy-making areas are re-defined as an increasing range of policy processes is drawn into agendas considered relevant for future action on climate change.

DOI: 10.4324/9781003452041-1

A second, related dynamic is a re-framing of policy processes dealing with climate change as a political issue. Partly in response to exogenous shocks such as the Covid pandemic and war in Ukraine (Anghel & Jones 2023), policies to achieve the target of decarbonization are defined as part of sustainable responses to economic, health and security crises in varying forms of conjunction with other policy priorities such as economic growth, social justice, or geopolitical security (Meckling & Strecker 2023). As a result, a greater degree of variation has emerged between agendas in which zero-carbon targets are integrated into policy processes including trade (Laurens et al. 2022) and industrial policy (Meckling & Allan 2020, Allan et al. 2021, Lewis 2021). Agendas have emerged that merge climate targets with economic and social policy issues; as a result, climate policy issues are linked and amalgamated with a wider and more varied set of policies.

An important variant of this development has been created through the launch of green industrial policy programs, adopted in both the EU and US particularly in response to the Covid-19 pandemic and war in Ukraine (Wendler 2023, Schramm et al. 2022, Massetti et al. 2022, Buti & Fabbrini 2023). Linking the expense of public funds in substantial size to the investment in zero-carbon technology and infrastructure, these programs promise to achieve a breakthrough for a clean energy transition and progress towards greater efficiency of energy use while re-affirming targets of economic growth and competitiveness. Particular attention has been drawn to policy developments in the US, where climate policies have been re-launched by the Biden administration after their decisive rollback during the previous Trump presidency (Jotzo et al. 2018, Kramer 2020, Mehling & Vishna 2017, Selby 2019). For the case of the EU, the promotion of green investment through expense from the Resilience and Recovery Facility (RRF) has created a new stage of climate action in terms of the expenditure raised and policy instruments applied in response to the Covid pandemic (Buti & Fabbrini 2023). While programs launched in the EU and US have distinctive differences that will be discussed in detail in the empirical part, a similarity of both is a re-direction of the policy-making approach to tackle climate change: namely, the departure from the use of restrictive regulation to reduce GHG emissions and the turn towards the provision of positive financial incentives for more carbon-efficient forms of production and consumption as a new variant of "green" industrial policy (Meckling & Strecker 2023, Meckling 2021, Newell et al. 2021, Nilsson et al. 2021, Pianta & Lucchese 2020).

Whether this shift signals a broadening or critical juncture for climate change governance remains to be seen: the current turn towards a more incentive-driven approach to climate action could be one that fits only to a limited set of settings and political contexts. These include situations of crisis management, or cases of climate politics where political pushback is felt against more stringent regulatory rules to cut carbon emissions. Evaluating whether the more incentive-driven approach to climate action signifies a general turn or episode is a major question for future research.

Aside from evaluating the substantive policy content of these packages, an important aspect of the re-framing of climate governance and creation of new linkages between policy fields is how it affects the politics of climate change. Decarbonization targets have become associated with a variety of policy-making processes and assume different degrees of visibility from a top priority to a more secondary aspect within these fields. This point is clarified by considering the different policy-making goals pursued under the heading of the green recovery. Beyond proclaimed targets of decarbonization, programs currently enacted under the REPowerEU (REPEU) plan and Inflation Reduction Act (IRA) in the US address a range of objectives, including: increased energy security and competitiveness in key energy transition industries, the creation of jobs with higher labor standards and help for regions affected by the exit from fossil fuel industries as well as targeted support for minority and indigenous communities (White House 2022, 2023, CRS 2022).

A major question for evaluating the politics of climate change in this context is whether it becomes more salient or is depoliticized as a political issue. Legislation considered as a breakthrough for climate action in the US has been launched as one primarily aiming at better infrastructure and less inflation; here, it appears that de-emphasizing climate targets in favor of promises of better jobs and growth has been applied to overcome opposition against policies explicitly aiming at curbing emissions to tackle climate change.

To summarize, the two dynamics discussed here – namely, the expansion and re-framing of governance processes associated with the complex issue of climate change – foreground questions about the interaction between its politics and policy-making dimension in a new context. We observe change in how climate action is defined and prioritized as a political issue, raising the question of how the effects of this change on subsequent policy-making results can be understood and theorized. In this regard, a key challenge for current research is to understand the shift of boundaries, institutions and decision-making processes that define processes of climate governance, and to evaluate them in terms of stability and change. This point touches on a major question of current research in this field: namely, to what degree efforts to achieve decarbonization require conditions of policy-making stability as a set of stable parameters concerning extant policy priorities, instruments and decision-making processes; or alternatively, to what degree action against climate change is advanced through disruptive and conflictual expansion and change (Paterson et al. 2022, Jordan & Moore 2020, Jordan et al. 2014). The changes brought about through these dynamics, and their impact on ongoing efforts to promote climate policies in the two cases of the EU and US, are the focus of this book.

1.1 Question and rationales of the book

The present analysis zooms in on a major component of the current expansion and re-framing of climate governance: namely, the adoption of green industrial policy agendas that combine the release of public funds for investment with conditionality rules and regulation requiring the promotion of climate-friendly technologies and forms of production and consumption. For the case of the EU, this analysis covers particularly the adoption of the NextGenerationEU (NGEU) and REPEU programs in conjunction with the climate governance framework of the EGD and as a major step towards its expansion. The case study on the US focuses on the adoption of public investment programs targeted at zero-carbon infrastructure and technology based on green conditionality as promoted through the Bipartisan Infrastructure Law (BIL) and particularly the IRA, as the latter is considered a major breakthrough towards a more ambitious climate policy in the US.

In its analysis of these policy packages, the present volume investigates the question: how do vectors associated with the concepts of policy-making stability and punctuation interact to create policy change in the evolution of climate governance frameworks in the EU and US? By relating political dynamics at the level of agenda-setting to questions of institutional and policy change, the overall aim of this volume is to contribute to research debates that seek to relate insights about the interaction of the politics and policy-making dimensions of action against climate change. More specifically, the rationales of this volume can be specified in three ways:

(1) The first rationale is descriptive: considering the still scarce literature on green industrial policy programs in the EU and US and even greater dearth of comparative research comparing both entities as contrasting cases of climate governance, our first aim is to understand the scope, direction and impact of change introduced by the green recovery programs covered here. To this end, the subsequent analysis will inquire how these policies introduce change to proclaimed agendas of climate action in both the EU and US, and to what degree they cause institutional and policy change.

(2) A second aim is analytical: the subsequent sections present a theoretical model based on punctuated equilibrium theory (PET) that helps to specify, evaluate and relate vectors of policy-making stability and change. To this end, the framework presented in the subsequent chapter specifies concepts and indicators that cover key stages of the policy process from agenda-setting to policy formulation and decision-making for the particular field of climate action. Here, the aim is to connect observations from the different stages of decision-making to identify dynamics of policy-making that can be classified as associated with either policy-making stability or punctuation.

(3) Finally, the present analysis seeks to make a broader theoretical contribution to the ongoing research debate discussing the role of factors for policy stability as opposed to sources of disruption, including events of exogenous shock and dynamics of political contestation, concerning their impact on the future development of climate change policy. A range of perspectives contribute to this debate, including those based on political economy, party political or discursive approaches; the contribution offered here is focused on policy beliefs and their relevance for shaping political agendas and subsequent interactions of policy venues and decision-making processes.

While this volume has an issue-specific focus on climate change policy in general and green recovery programs in particular, its puzzle and question relates to a classic debate within the policy-making literature: namely, how political agendas form in response to shifts of political attention and events of exogenous shock, and how the recasting of political problems through linkages with other issues affects subsequent decision-making venues, processes and eventual results (Baumgartner et al. 2009, 2018).

The main argument developed through the subsequent analysis is that research approaching climate change as an issue of political contestation and conflict should not miss its variability as a set of demands for policy change, and the relevance of shifts in the boundaries within which it is proclaimed and negotiated as a problem for society. Setting these boundaries, and achieving their stability after a phase of policy creation, may have become one of the most important factors for promoting climate action against opposed interests.

1.2 State of the art: climate governance, policy stability and political conflict

The present volume speaks to several research literatures that seek to relate observations about the politics of climate change – particularly at the level of agenda-setting, legislative negotiation and political contestation – with its policy dimension in terms of actual progress towards zero-carbon targets. More specifically, the subsequent analysis builds on three main literatures: (1) policy stability, politicization and political conflict in climate governance; (2) climate policy integration (CPI) and theories of the policy process; and (3) comparative perspectives on climate policy in the EU and US.

(1) Policy stability, politicization and political conflict in climate
     governance

First, the present volume adds to a growing literature that investigates the effects of political conflict and disruption as a factor for progress of action against climate change in comparison to conditions of stability and continuity. This debate is brought into sharp focus in a recent forum article opposing policy stability and re-politicization as antagonistic concepts at the levels of both actual climate politics and its theoretical evaluation (Paterson et al. 2022). Related literatures inquire whether increased political conflict and politicization are "good" or "bad" for progress against climate change (Pepermans & Maeseele 2016). While the dualism of stability and conflict lumps together a range of arguments from different literatures and raises some questions about conceptual clarity, it is productive for distinguishing between two sets of perspectives on policy processes dealing with climate change: positions associated with policy stability make the case that a stable set of agendas, institutions and decision-making processes is required to enact a long-term trajectory of policies to achieve targets of carbon neutrality; a related point is that legally binding long-term commitments to decarbonization particularly through climate laws, a strong role of scientific expertise, and remoteness from majoritarian politics and polarized debate support policy development along these lines (Jordan & Moore 2020, Delbeke & Vis 2019, Carlson & Burtraw 2019).

Applied to empirical cases, it seems that particularly the EU, as a highly compound political system that is largely removed from strong dynamics of adversarial politics, has followed a relatively long and continuous trajectory of climate policies that appears as a likely case of policy stability. In this regard, research covering the evolution of EU energy and climate governance has identified a gradual increase in the scope, density and strength particularly of regulatory policies (Oberthür & von Homeyer 2023, Skjaerseth 2021, Delbeke & Vis 2015, Oberthür & Dupont 2015), relatively successful mechanisms of CPI (Dupont 2016), a leadership role of the Commission for continuing its climate agenda over several electoral cycles (Rietig & Dupont 2021), and the foundation of EU climate action in a relatively stable and comprehensive political framing culminating in the EGD agenda (Wendler 2022a). An open question is whether exogenous shocks such as the Covid pandemic or war in Ukraine have brought aspects of disruption into this development. In this regard, extant research seems to indicate that even in a context of turbulence and crisis, climate policy development in the EU remains relatively stable with regard to the priority of climate action (Wendler 2023, von Homeyer et al. 2021, Siddi 2021, Rietig 2021), with a greater impact observed on environmental policy (Burns & Tobin 2016).

By contrast, a second strand of the literature associates dynamics of political conflict and subsequent disruption of extant policy-making coalitions with progress against climate change. An important part of this literature is explicitly normative by calling for more intense democratic debate about the issue of climate change, either in mass publics (Willis 2020) or in more networked deliberative settings (Stevenson & Dryzek 2014). Other contributions take a critical perspective by questioning the post-political consensus, particularly on the paradigm of economic growth within the context of climate action agendas (Blühdorn & Deflorian 2021, Kenis & Mathijs 2014, Ajl 2021). Arguments in favor of more intense politicization of climate change as a political issue are also made on empirical grounds: primarily, that stronger opposition to status quo policies will create pressure on policy-makers to adopt targets of decarbonization and help to achieve breakthroughs against the resistance of vested interests, particularly of the fossil fuel industry. Empirical evidence for such dynamics, however, remains mixed; particularly for the case of the US, where climate change is a deeply polarized and highly partisan issue, extant studies indicate a primarily destructive effect of intense political conflict on policy-making (Aklin & Mildenberger 2020, Mildenberger 2020, Mann 2021). Against this background, it seems relevant to inquire whether the approach of the Biden administration to merge climate targets with objectives of middle-class prosperity and growth while reducing the salience especially of restrictive measures to curb carbon emissions is more successful than previous attempts to adopt regulation at the federal level (Bang 2021).

A combined perspective on the respective relevance and interaction of stability and political conflict is created by the growing literature on the politicization of climate change as a topic of political mobilization and contestation (Zürn 2019, Davies et al. 2021, McCright & Dunlap 2011; for a literature review cp. Wendler 2022b: 3–5). Here, a range of contributions have documented the increased salience and contentiousness of climate-related topics at the level of party politics (Carter et al. 2018, Carter & Little 2020, Farstad 2018, Gustafson et al. 2019, Båtstrand 2015) and particularly through climate denial (Fischer 2019) and backlash (Patterson 2023, Stokes 2016), the role of populist right parties (Marquardt & Lederer 2022, Huber et al. 2021, Almiron et al. 2019, Cann & Raymond 2018, Lockwood 2018, Swyngedouw 2022), civil society mobilization (Berglund & Schmidt 2020, Corry & Reiner 2021), as a subject of debate within established and new social media (Chinn et al. 2020, Bolsen & Shapiro 2018, Boykoff 2011, Zhou 2016) and within parliamentary settings such as the European Parliament (EP) (Kinski & Servent 2022, Wendler 2019). An open question for future research, however, is the question of how to capture climate change as an issue in given empirical material such as party manifestos or parliamentary speech in conjunction with other policy issues. In this regard, challenges arise particularly when climate issues are amalgamated with economic and social ones through the dynamics described above, where they are no longer adequately identified as a component or even subfield of environmental politics. How the "climateness" of topics – their relative proximity to or linkage with ambitions of decarbonization – affects their political contestation is a challenging issue for both qualitative research and text-as-data research (Grimmer 2022).

In summation, extant empirical research appears inconclusive with regard to the question whether more intense political conflict and disruptive events work as a prompt or a barrier for breakthroughs towards more progressive climate policies. Applicable theoretical approaches offer different assumptions in this regard: for the particular case of EU multi-level governance, post-functionalist theory has established itself as the default approach for linking increased politicization with constraints on decision-making; its arguments are based on the entry of previously technocratic issues into arenas of mass politics and the cultural-identitarian framing of counter-arguments by populist political entrepreneurs (Hooghe & Marks 2019). It is evident that the issue of climate change as adopted through the EGD agenda would offer itself to attempts of counter-mobilization based on typical populist rhetoric against detached elites. Based on this approach, we would therefore almost certainly expect increasing political blockades against components of the EGD agenda as a result of increasing politicization, even if empirical support for the constraining dissensus predicted by post-functionalism remains mixed (Börzel & Risse 2018). In a contrary perspective, major theories of the policy process associate sources of political conflict and disruption – caused by exogenous shocks as well as shifts of attention by political publics and major decision-makers – with a shake-up of policy monopolies, creation of new actor coalitions and subsequent breakthroughs to policy change (Weible & Workman 2022, Weible & Sabatier 2017).

The review of the research debate sketched here shows that the dualism of policy stability versus disruption is relevant for the book's focus but also needs further refinement in order to be applicable to the subsequent analysis. In this sense, "policy stability" is a broad term that refers to two distinct analytical perspectives: first, the politics dimension of climate governance, and particularly the continuity of institutional and political conditions of the policy process and its distance from polarized contestation; and its policy dimension, defined by the successful adoption of incremental and cumulative policy change in favor of decarbonization. Both disruptive breakthroughs towards stringent climate action and policy reversals fall outside this category, leaving policy stability as a problematic concept without further clarification of the degree and directionality of policy change.

Similarly, politicization appears as a relatively rough concept to be applied to the issue of climate change when considering that climate-related issues are raised in a wide variety of contexts with different scope and often amalgamated with other economic or social issues. In this sense, it appears simplistic to simply distinguish between "more" or "less" politicized debate concerning the challenge of climate change. Considering the expansion and diffusion of climate targets into a broad range of areas as discussed above, a more nuanced understanding is required for capturing the different degrees of density and priority through which climate-related demands are raised in different frameworks of policy-making. A key political change in the more recent development of the politics of climate change, in this sense, may well lie not in the increase of political conflict within a given set of policies and institutions, but in a shift of the boundaries within which priorities of climate action are proposed, negotiated and contested.

(2) Climate policy integration and theories of the policy process

Particularly the latter point creates a connection between the analysis of climate politics and a related research debate: namely, how zero-carbon targets are transferred to policy-making areas not a priori associated with climate change, and what factors matter for the success of these transfers. CPI is the most relevant concept in this regard, and is broadly defined as a process for "the incorporation of the aims of climate change mitigation and adaptation into all stages of policy-making in other policy sectors" (Mickwitz et al. 2009). While the concept is relevant for the topics and questions discussed in this volume, several open questions from the related research debate are raised about its conceptualization in the light of policy developments addressed here.

The first of these is at a simple conceptual level and concerns the distinction between climate and "other" policy fields; many definitions rely on this distinction to describe the expansion of climate governance processes beyond a core or original field. Considering the comprehensive scope of recent climate agendas such as the EGD as a cross-government issue, however, the distinction between a core of climate policy and some external realm evidently makes little sense, while glossing over remaining asymmetries between the strengths of linkages of different sectors with core climate targets. By contrast, it could be asked what remains of climate policy if CPI is removed, leaving no obvious answers where its original format as a field of policy-making can be found. In this sense, it has been asked what the distinction is between CPI and climate policy as such (Adelle & Russel 2013: 2).

The point to be made here is that it seems increasingly problematic to assume the existence of a stable or predefined political and institutional core of climate policy from which ambitions or policies are exported to other fields. By contrast, recent developments seem to underline the fact that climate policy is established through the application of a set of core demands for transformative change – primarily, the reduction of GHG emissions – to a range of extant policy-making fields whose selection and interlinkage are fundamentally variable. More recent definitions resonate with this understanding by defining CPI as an "increasingly ambitious integration of policy goals, governance arrangements and policy processes related to climate change adaptation and mitigation" (von Lüpke & Well 2020: 834). This aspect leads to the related point of whether CPI should be distinguished from the older concept of environmental policy integration (Lenschow & Gravey 2021, Adelle & Russel 2013). Resonating with the understanding of climate policy proposed here, most contributions comparing the two concepts understand CPI as related but distinct and potentially broader than EPI (Rietig 2019: 230). Furthermore, recent policy developments associated with the energy transition – particularly the issue of siting and permitting, use of nuclear energy or extraction of critical resources for battery storage – raise questions about the competition or even conflict between climate and classical environmental priorities as a new topic for research.

Finally, several points in the previous discussion of this chapter lead to an insight of more general importance: namely, that the linkage of climate targets with other policy priorities concerns both the politics dimension of climate governance and its policy-making dimension, as well as their mutual inter-relation. In this context, one of the puzzles identified at the outset was how new forms of integration and amalgamation of climate targets in political agendas affect subsequent policy-making processes and their institutional foundation.

This more comprehensive approach to the idea of CPI is supported by more recent, primarily process-based definitions of CPI in the literature that trace its evolution across various stages of the policy process (Cejudo & Trein 2023a,b). In this sense, CPI has been recast as the result of interest-driven and often conflictual political agency, and conceptualized through the interaction of four dimensions spanning the entire policy process from its framing to the choice of policy subsystems, definition of policy goals and choice and instruments (Candel & Biesbroek 2016, Biesbroek & Candel 2020). The proximity of this approach to theories and concepts of the policy process is obvious (Weible & Sabatier 2017, Workman et al. 2022). A strength of this approach, therefore, is that it opens up a theoretical perspective to discuss CPI as a dynamic at work in both the politics and policy dimensions of climate governance processes, and to evaluate the interactions, tensions and asymmetries between them. This conceptualization is reflected in the theoretical framework presented in the subsequent chapter.

(3) Comparative perspectives on climate policy in the EU and US

Finally, this book is embedded in the comparative literature on the EU and US as two cases of climate policy-making. These two entities are relevant cases with regard to their roles both as the two major GHG emitters within the highly industrialized world and as agents of climate negotiations at the global level (Luterbacher & Sprinz 2018). Nevertheless, direct comparisons between both systems with regard to climate policy are still relatively scarce (cp. Wendler 2022a, Skjaerseth et al. 2013, Carlarne 2010). Extant comparative surveys that include the EU and US either as separate case studies (Wurzel et al. 2021) or as two cases in a broader global comparison (Jahn 2016) tend to portray the two, respectively, as a relative leader and laggard with clearly contrasting records of climate policy-making (Wurzel et al. 2017, Rayner & Jordan 2013). This point is not surprising given differences between the EU and US concerning their participation in global climate agreements, the adoption of regulatory tools for the restriction of carbon emissions such as emissions trading or green energy standards, and the per capita emission of GHGs as well as their reduction over time (Boasson & Wettestad 2013, Brewer 2015, Kalantzos 2017). Among the explanations for the discrepancy between the climate policies of these two political systems, arguments concerning the politics of climate change and organized climate denial particularly stand out (Skjaerseth et al. 2013). These are documented in a range of studies covering political polarization between the two major parties in the US (Gustafson et al. 2019), the politics of the US Congress (Guber et al. 2021) and media coverage of the topic (Chinn et al. 2020). These contrasts have been brought to a head during the 2017–21 Trump presidency and particularly the announcement of withdrawal by the US from the Paris Agreement (Jotzo et al. 2018).

The policy-making developments identified at the outset of this chapter cast a new light on this comparison, in two respects. One is that the programs subsumed under the concept of green industrial policy share some broad similarities between the EU and US, with regard to both the overall size of funds mobilized and the general approach of combining subsidies and incentives for investment with conditionality criteria tied to reduction of GHG emissions. From this point of departure, an interesting aspect of comparison is to discuss distinctive differences between the two sets of programs; these include not only their different use of market-based processes and state-led investment but also their different degrees of foundation in a regulatory framework to prescribe reductions of GHG emissions (Meckling & Strecker 2023). In this sense, the comparison between green industrial programs in the EU and US raises the question whether they introduce aspects of convergence or in fact reinforce the different approaches of both systems to policy-making on the problem of climate change.

A second aspect concerns questions about how the adoption of green industrial programs affects the interaction between the EU and US: particularly, how they contribute to the noticeable general turn of climate governance towards a more geopolitical perspective focused on securing supply chains, competitiveness and energy security (Allan 2021). So far, it seems that legislation adopted in the US has a more strategic orientation and stronger impact at this level than comparable programs in the EU; however, responses from the European side in this regard have been debated for some time, especially through the proposal of a Green Deal Industrial Plan (GDIP) and related initiatives (European Commission 2023). Seen from this perspective, a comparative perspective on green industrial policies of the EU and US and their interaction seems even more pertinent than before.

To summarize, the subsequent comparative analysis of climate and green industrial policies in the EU and US covers policy-specific developments but is set at the intersection of debates and literatures that reach beyond the specific case of climate change policy. It is clear that some of the broader questions for research discussed in this section are beyond the scope of the present volume. Its main rationale, however, is to provide an impulse to a key research debate about climate politics and policy-making: namely, how to conceptualize opposing vectors of policy-making stability and punctuation as factors for progress of policy-making against climate change.

1.3 Outline and main argument of the book

The subsequent analysis is presented in four steps. First, Chapter 2 presents the theoretical framework of the analysis, building on PET to discuss the interaction between evolving agendas and institutional venues of climate change governance, and model their subsequent effects on policy outputs in terms of positive and negative feedback. Second, Chapters 3 and 4 apply this model to the analysis of recent policy-making developments in the EU and US and, more specifically, their enactment through the NGEU and REPEU packages in the case of the EU, and the BIL and IRA in the US. The comparison will show the contrast between developments primarily characterized by policy stability and continuity in the EU as compared to greater degrees of political conflict, disruptive policy change and uncertainty in the US. The conclusion summarizes the main findings from these two case studies and provides an outlook on ongoing policy developments and perspectives for future research (Chapter 5).

References

Adelle, Camilla, and Duncan Russel. 2013. "Climate Policy Integration: A Case of Déjà Vu?" *Environmental Policy and Governance* 23(1): 1–12. doi:10.1002/eet.1601.

Ajl, Max. 2021. *A People's Green New Deal*. London: Pluto Press.

Aklin, Michaël, and Matto Mildenberger. 2020. "Prisoners of the Wrong Dilemma: Why Distributive Conflict, Not Collective Action, Characterizes the Politics of Climate Change." *Global Environmental Politics* 20(4): 4–27. doi:10.1162/glep_a_00578.

Allan, Bentley, Joanna I. Lewis, and Thomas Oatley. 2021. "Green Industrial Policy and the Global Transformation of Climate Politics." *Global Environmental Politics* 21(4): 1–19. doi:10.1162/glep_a_00640.

Almiron, Núria, and Jordi Xifra i Triadú. 2019. *Climate Change Denial and Public Relations: Strategic Communication and Interest Groups in Climate Inaction.* Abingdon, OX; New York, NY: Routledge.

Anghel, Veronica, and Erik Jones. 2023. "Is Europe Really Forged Through Crisis? Pandemic EU and the Russia – Ukraine War." *Journal of European Public Policy* 30(4): 766–86. doi:10.1080/13501763.2022.2140820.

Aronoff, Kate, Alyssa Battistoni, Daniel Aldana Cohen, and Thea Riofrancos. 2019. *A Planet to Win. Why We Need a Green New Deal.* London, New York: Verso.

Bang, Guri. 2021. "The United States: Conditions for Accelerating Decarbonisation in a Politically Divided Country." *International Environmental Agreements: Politics, Law and Economics* 21(1): 43–58. doi:10.1007/s10784-021-09530-x.

Båtstrand, Sondre. 2015. "More than Markets: A Comparative Study of Nine Conservative Parties on Climate Change." *Politics & Policy* 43(4): 538–61. doi:10.1111/polp.12122.

Baumgartner, Frank, and Bryan Jones. 2015. *The Politics of Information. Problem Definition and the Course of Public Policy in America.* Chicago: University of Chicago Press.

Baumgartner, Frank, Bryan Jones, and Peter B. Mortensen. 2018. "Punctuated Equilibrium Theory: Explaining Stability and Change in Public Policymaking." In *Theories of the Policy Process*, eds. Christopher M. Weible and Paul A. Sabatier. New York, NY: Westview Press, 55–102.

Baumgartner, Frank, Christian Breunig, Green Pedersen, Bryan Jones, Peter B. Mortensen, Michiel Nuytemans, and Stefaan Walgrave. 2009. "Punctuated Equilibrium in Comparative Perspective." *American Journal of Political Science* 53(3): 603–20.

Berglund, Oscar, and Daniel Schmidt. 2020. *Extinction Rebellion and Climate Change Activism: Breaking the Law to Change the World.* Basingstoke: Palgrave Macmillan.

Biesbroek, Robbert, and Jeroen J. L. Candel. 2020. "Mechanisms for Policy (Dis)Integration: Explaining Food Policy and Climate Change Adaptation Policy in the Netherlands." *Policy Sciences* 53(1): 61–84.

Bloomfield, Jon, and Fred Steward. 2020. "The Politics of the Green New Deal." *The Political Quarterly* 91(4): 770–79. doi:10.1111/1467-923X.12917.

Blühdorn, Ingolfur, and Michael Deflorian. 2021. "Politicisation Beyond Post-Politics: New Social Activism and the Reconfiguration of Political Discourse." *Social Movement Studies* 20(3): 259–75. doi:10.1080/14742837.2021.1872375.

Boasson, Elin Lerum, and Jørgen Wettestad. 2013. *EU Climate Policy: Industry, Policy Interaction and External Environment.* Farnham [u.a.]: Ashgate.

Bolsen, Toby, and Matthew A. Shapiro. 2018. "The US News Media, Polarization on Climate Change, and Pathways to Effective Communication." *Environmental Communication* 12(2): 149–63. doi:10.1080/17524032.2017.1397039.

Bolsen, Toby, Risa Palm, and Justin T. Kingsland. 2019. "Counteracting Climate Science Politicization with Effective Frames and Imagery." *Science Communication* 41(2): 147–71. doi:10.1177/1075547019834565.

Bongardt, Annette, and Francisco Torres. 2022. "The European Green Deal: More than an Exit Strategy to the Pandemic Crisis, a Building Block of a Sustainable European Economic Model*." *JCMS: Journal of Common Market Studies* 60(1): 170–85. doi:10.1111/jcms.13264.

Börzel, Tanja A., and Thomas Risse. 2018. "From the Euro to the Schengen Crises: European Integration Theories, Politicization, and Identity Politics." *Journal of European Public Policy* 25(1): 83–108. doi:10.1080/13501763.2017.1310281.

Boykoff, Maxwell T. 2011. *Who Speaks for the Climate?: Making Sense of Media Reporting on Climate Change.* Cambridge: Cambridge University Press.

Bressanelli, Edoardo, Christel Koop, and Christine Reh. 2016. "The Impact of Informalisation: Early Agreements and Voting Cohesion in the European Parliament." *European Union Politics* 17(1): 91–113. doi:10.1177/1465116515608704.

Brewer, Thomas L. 2015. *The United States in a Warming World: The Political Economy of Government, Business, and Public Responses to Climate Change.* Cambridge: Cambridge University Press.

Burns, Charlotte, and Paul Tobin. 2016. "The Impact of the Economic Crisis on European Union Environmental Policy." *JCMS: Journal of Common Market Studies* 54(6): 1485–94. doi:10.1111/jcms.12396.

Buti, Marco, and Sergio Fabbrini. 2023. "Next Generation EU and the Future of Economic Governance: Towards a Paradigm Change or Just a Big One-Off?" *Journal of European Public Policy* 30(4): 676–95. doi:10.1080/13501763.2022.2141303.

Candel, Jeroen J. L., and Robbert Biesbroek. 2016. "Toward a Processual Understanding of Policy Integration." *Policy Sciences* 49(3): 211–31. doi:10.1007/s11077-016-9248-y.

Cann, Heather W., and Leigh Raymond. 2018. "Does Climate Denialism Still Matter? The Prevalence of Alternative Frames in Opposition to Climate Policy." *Environmental Politics* 27(3): 433–54. doi:10.1080/09644016.2018.1439353.

Carlarne, Cinnamon Piñon. 2010. *Climate Change Law and Policy: EU and US Approaches.* Oxford [u.a.]: Oxford University Press.

Carlson, Ann, and Dallas Burtraw, eds. 2019. *Lessons from the Clean Air Act. Building Durability and Adaptability into U.S. Climate and Energy Policy.* Cambridge: Cambridge University Press.

Carter, Neil, and Conor Little. 2020. "Party Competition on Climate Policy: The Roles of Interest Groups, Ideology and Challenger Parties in the UK and Ireland." *International Political Science Review* 42(1). doi:10.1177/0192512120972582.

Carter, Neil, Robert Ladrech, Conor Little, and Vasiliki Tsagkroni. 2018. "Political Parties and Climate Policy: A New Approach to Measuring Parties' Climate Policy Preferences." *Party Politics* 24(6): 731–42. doi:10.1177/1354068817697630.

Cejudo, Guillermo M., and Philipp Trein. 2023a. "Pathways to Policy Integration: A Subsystem Approach." *Policy Sciences* 56(1): 9–27. doi:10.1007/s11077-022-09483-1.

Cejudo, Guillermo M., and Philipp Trein. 2023b. "Policy Integration as a Political Process." *Policy Sciences* 56(1): 3–8. doi:10.1007/s11077-023-09494-6.

Chinn, Sedona, P. Sol Hart, and Stuart Soroka. 2020. "Politicization and Polarization in Climate Change News Content, 1985–2017." *Science Communication* 42(1): 112–29. doi:10.1177/1075547019900290.

Corry, Olaf, and David Reiner. 2021. "Protests and Policies: How Radical Social Movement Activists Engage with Climate Policy Dilemmas." *Sociology* 55(1): 197–217. doi:10.1177/0038038520943107.

CRS (Congressional Research Service). 2022. "Inflation Reduction Act of 2022 (IRA): Provisions Related to Climate Change." https://crsreports. congress.gov/product/pdf/R/R47262#:~:text=IRA%20contains%20ei ght%20titles%2C%20each,resilience%20to%20climate%20change%20 impacts.

Dalby, Simon, and Shannon O'Lear. 2016. *Reframing Climate Change: Constructing Ecological Geopolitics*. Abingdon, OX; New York, NY: Routledge.

Davies, Anna R., Vanesa Castán Broto, and Stephan Hügel. 2021. "Editorial: Is There a New Climate Politics?" *Politics and Governance* 9(2): 1–7. doi:10.17645/pag.v9i2.4341.

Delbeke, Jos, and Peter Vis. 2015. *EU Climate Policy Explained*. Abingdon, OX [u.a.]: Routledge.

Delbeke, Jos, and Peter Vis. 2019. *Towards a Climate-Neutral Europe: Curbing the Trend*. London: Routledge.

Dirikx, Astrid, and Dave Gelders. 2010. "To Frame Is to Explain: A Deductive Frame-Analysis of Dutch and French Climate Change Coverage During the Annual UN Conferences of the Parties." *Public Understanding of Science* 19(6): 732–42. doi:10.1177/0963662509352044.

Domorenok, Ekaterina, and Paolo Graziano. 2023. "Understanding the European Green Deal: A Narrative Policy Framework Approach." *European Policy Analysis* 9(1): 9–29. doi:10.1002/epa2.1168.

Dupont, Claire. 2016. *Climate Policy Integration into EU Energy Policy*. Abingdon, OX [u.a.]: Routledge.

Eckert, Sandra. 2021. "The European Green Deal and the EU's Regulatory Power in Times of Crisis." *JCMS: Journal of Common Market Studies* 59(S1): 81–91. doi:10.1111/jcms.13241.

European Commission. 2023. "A Green Deal Industrial Plan for the Net-Zero Age." https://eur-lex.europa.eu/legal-content/EN/TXT/PDF/?uri= CELEX:52023DC0062.

Farstad, Fay M. 2018. "What Explains Variation in Parties' Climate Change Salience?" *Party Politics* 24(6): 698–707. doi:10.1177/1354068817693473.

Fischer, Frank. 2019. "Knowledge Politics and Post-Truth in Climate Denial: On the Social Construction of Alternative Facts." *Critical Policy Studies* 13(2): 133–52. doi:10.1080/19460171.2019.1602067.

Grasso, Marco, and Ezra M. Markowitz. 2015. "The Moral Complexity of Climate Change and the Need for a Multidisciplinary Perspective on Climate Ethics." *Climatic Change* 130(3): 327–34. doi:10.1007/s10584-014-1323-9.

Grimmer, Justin. 2022. *Text as Data a New Framework for Machine Learning and the Social Sciences.* Princeton, NJ: Princeton University Press.

Guber, Deborah Lynn, Jeremiah Bohr, and Riley E. Dunlap. 2021. "'Time to Wake Up': Climate Change Advocacy in a Polarized Congress, 1996–2015." *Environmental Politics* 30(4): 538–58. doi:10.1080/09644016.2020.1786333.

Gustafson, Abel, Seth A. Rosenthal, Matthew T. Ballew, Matthew H. Goldberg, Parrish Bergquist, John E. Kotcher, Edward W. Maibach, and Anthony Leiserowitz. 2019. "The Development of Partisan Polarization over the Green New Deal." *Nature Climate Change* 9: 940–44. doi:10.1038/s41558-019-0621-7.

Hainsch, Karlo, Konstantin Löffler, Thorsten Burandt, Hans Auer, Pedro Crespo del Granado, Paolo Pisciella, and Sebastian Zwickl-Bernhard. 2022. "Energy Transition Scenarios: What Policies, Societal Attitudes, and Technology Developments Will Realize the EU Green Deal?" *Energy* 239: 122067. doi:10.1016/j.energy.2021.122067.

Hayes, Jarrod, and Janelle Knox-Hayes. 2014. "Security in Climate Change Discourse: Analyzing the Divergence Between US and EU Approaches to Policy." *Global Environmental Politics* 14(2): 82–101. doi:10.1162/GLEP_a_00230.

Hickel, Jason, and Giorgos Kallis. 2020. "Is Green Growth Possible?" *New Political Economy* 25(4): 469–86. doi:10.1080/13563467.2019.1598964.

von Homeyer, Ingmar, Sebastian Oberthür, and Andrew J. Jordan. 2021. "EU Climate and Energy Governance in Times of Crisis: Towards a New Agenda." *Journal of European Public Policy* 28(7): 959–79. doi:10.1080/13501763.2021.1918221.

Hooghe, Liesbet, and Gary Marks. 2009. "A Postfunctionalist Theory of European Integration: From Permissive Consensus to Constraining

Dissensus." *British Journal of Political Science* 39(1): 1–23. doi:10.1017/ S0007123408000409.

Hooghe, Liesbet, and Gary Marks. 2019. "Grand Theories of European Integration in the Twenty-First Century." *Journal of European Public Policy* 26(8): 1113–33. doi:10.1080/13501763.2019.1569711.

Huber, Robert A., Tomas Maltby, Kacper Szulecki, and Stefan Ćetković. 2021. "Is Populism a Challenge to European Energy and Climate Policy? Empirical Evidence Across Varieties of Populism." *Journal of European Public Policy* 28(7): 998–1017. doi:10.1080/13501763.2021.1918214.

Incropera, Frank. 2016. *Climate Change: A Wicked Problem. Complexity and Uncertainty at the Intersection of Science, Economics, Politics, and Human Behavior.* Cambridge: Cambridge University Press.

Jahn, Detlef. 2016. *The Politics of Environmental Performance: Institutions and Preferences in Industrialized Democracies.* Cambridge: Cambridge University Press.

Jordan, Andrew, and Brendan Moore. 2020. *Durable by Design? Policy Feedback in a Changing Climate.* Cambridge: Cambridge University Press.

Jordan, Andrew, and Elah Matt. 2014. "Designing Policies That Intentionally Stick: Policy Feedback in a Changing Climate." *Policy Sciences* 47(3): 227–47.

Jotzo, Frank, Joanna Depledge, and Harald Winkler. 2018. "US and International Climate Policy Under President Trump." *Climate Policy* 18(7): 813–17.

Kalantzakos, Sophia. 2017. *The EU, US and China Tackling Climate Change: Policies and Alliance for the Anthropocene.* London; New York: Routledge.

Kenis, A., and E. Mathijs. 2014. "Climate Change and Post-Politics: Repoliticizing the Present by Imagining the Future?" doi:10.1016/ J.GEOFORUM.2014.01.009.

Kinski, Lucy, and Ariadna Ripoll Servent. 2022. "Framing Climate Policy Ambition in the European Parliament." *Politics and Governance* 10(3): 251–63. doi:10.17645/pag.v10i3.5479.

Kramer, Ronald. 2020. "Rolling Back Climate Regulation: Trump's Assault on the Planet." *Journal of White Collar and Corporate Crime* 1(2): 123–30.

Kreppel, Amie. 2018. "Bicameralism and the Balance of Power in EU Legislative Politics." *Journal of Legislative Studies* 24(1): 11–33. doi:10.1080/13572334.2018.1444623.

Latour, Bruno. 2018. *Down to Earth. Politics in the New Climatic Regime.* Cambridge: Polity Press.

Laurens, Noémie, Clara Brandi, and Jean-Frédéric Morin. 2022. "Climate and Trade Policies: From Silos to Integration." *Climate Policy* 22(2): 248–53. doi:10.1080/14693062.2021.2009433.

Lenschow, Andrea, and Viviane Gravey, eds. 2021. *Environmental Policy in the EU: Actors, Institutions and Processes.* Abingdon, OX [u.a.]: Routledge.

www.routledge.com/Environmental-Policy-in-the-EU-Actors-Instituti ons-and-Processes/Jordan-Gravey/p/book/9781138392168   (February 22, 2024).

Levin, Kelly, Benjamin Cashore, Steven Bernstein, and Graeme Auld. 2012. "Overcoming the Tragedy of Super Wicked Problems: Constraining Our Future Selves to Ameliorate Global Climate Change." *Policy Sciences* 45(2): 123–52. doi:10.1007/s11077-012-9151-0.

Lewis, Joanna I. 2021. "Green Industrial Policy After Paris: Renewable Energy Policy Measures and Climate Goals." *Global Environmental Politics* 21(4): 42–63. doi:10.1162/glep_a_00636.

Lockwood, Matthew. 2018. "Right-Wing Populism and the Climate Change Agenda: Exploring the Linkages." *Environmental Politics* 27(4): 712–32. doi:10.1080/09644016.2018.1458411.

von Lüpke, Heiner, and Mareike Well. 2020. "Analyzing Climate and Energy Policy Integration: The Case of the Mexican Energy Transition." *Climate Policy* 20(7): 832–45. doi:10.1080/14693062.2019.1648236.

Luterbacher, Urs, and Detlef F. Sprinz, eds. 2018. *Global Climate Policy: Actors, Concepts, and Enduring Challenges.* Cambridge, Massachusetts: The MIT Press.

Mann, Michael E. 2021. *The New Climate War: The Fight to Take Back Our Planet.* New York: Public Affairs, Hachette.

Marquardt, Jens, and Markus Lederer. 2022. "Politicizing Climate Change in Times of Populism: An Introduction." *Environmental Politics* 31(5): 735–54. doi:10.1080/09644016.2022.2083478.

Massetti, Emanuele, and Theofanis Exadaktylos. 2022. "From Crisis to Crisis: The EU in Between the Covid, Energy and Inflation Crises (and War)." *JCMS: Journal of Common Market Studies* 60(S1): 5–11. doi:10.1111/jcms.13435.

Mastini, Riccardo, Giorgos Kallis, and Jason Hickel. 2021. "A Green New Deal Without Growth?" *Ecological Economics* 179: 106832. doi:10.1016/j.ecolecon.2020.106832.

McCright, Aaron M., and Riley E. Dunlap. 2011. "The Politicization of Climate Change and Polarization in the American Public's Views of Global Warming, 2001–2010." *Sociological Quarterly* 52(2): 155–94. doi:10.1111/j.1533-8525.2011.01198.x.

McDonald, Matt. 2021. *Ecological Security: Climate Change and the Construction of Security.* Cambridge, New York: Cambridge University Press.

Meckling, Jonas. 2021. "Making Industrial Policy Work for Decarbonization." *Global Environmental Politics* 21(4): 134–47. doi:10.1162/glep_a_00624.

Meckling, Jonas, and Bentley B. Allan. 2020. "The Evolution of Ideas in Global Climate Policy." *Nature Climate Change* 10(5): 434–38. doi:10.1038/s41558-020-0739-7.

Meckling, Jonas, and Jesse Strecker. 2023. "Green Bargains: Leveraging Public Investment to Advance Climate Regulation." *Climate Policy* 23(4): 418–29. doi:10.1080/14693062.2022.2149452.

Mehling, Michael, and Antto Vihma. 2017. " 'Mourning for America'. Donald Trump's Climate Change Policy." www.fiia.fi/wp-content/uploads/2017/10/; analysis8_mourning_for_america-2.pdf.

Mickwitz, Per, Francisco Aix, Silke Beck, David Carss, and Nils Ferrand. 2009. *Climate Policy Integration, Coherence and Governance.* Helsinki: PEER. https://pure.au.dk/ws/files/56076592/PEER_Report2.pdf.

Mildenberger, Matto. 2020. *Carbon Captured: How Business and Labor Control Climate Politics.* Cambridge, Massachusetts: The MIT Press.

Newell, Peter, Matthew Paterson, and Martin Craig. 2021. "The Politics of Green Transformations: An Introduction to the Special Section." *New Political Economy* 26(6): 903–6. doi:10.1080/13563467.2020.1810215.

Nilsson, Lars J., Fredric Bauer, Max Åhman, Fredrik N. G. Andersson, Chris Bataille, Stephane de la Rue du Can, Karin Ericsson, et al. 2021. "An Industrial Policy Framework for Transforming Energy and Emissions Intensive Industries Towards Zero Emissions." *Climate Policy* 21(8): 1053–65. doi:10.1080/14693062.2021.1957665.

Oberthür, Sebastian, and Claire Dupont. 2015. *Decarbonization in the European Union: Internal Policies and External Strategies.* Basingstoke: Palgrave Macmillan.

Oberthür, Sebastian, and Claire Dupont. 2021. "The European Union's International Climate Leadership: Towards a Grand Climate Strategy?" *Journal of European Public Policy* 28(7): 1095–114. doi:10.1080/13501763.2021.1918218.

Oberthür, Sebastian, and Ingmar von Homeyer. 2023. "From Emissions Trading to the European Green Deal: The Evolution of the Climate Policy Mix and Climate Policy Integration in the EU." *Journal of European Public Policy* 30(3): 445–68. doi:10.1080/13501763.2022.2120528.

Ossewaarde, Marinus, and Roshnee Ossewaarde-Lowtoo. 2020. "The EU's Green Deal: A Third Alternative to Green Growth and Degrowth?" *Sustainability* 12(23): 9825. doi:10.3390/su12239825.

Paterson, Matthew. 2021a. *In Search of Climate Politics.* Cambridge: Cambridge University Press.

Paterson, Matthew. 2021b. " 'The End of the Fossil Fuel Age'? Discourse Politics and Climate Change Political Economy." *New Political Economy* 26(6): 923–36. doi:10.1080/13563467.2020.1810218.

Paterson, Matthew, Paul Tobin, and Stacy D. VanDeveer. 2022. "Climate Governance Antagonisms: Policy Stability and Repoliticization." *Global Environmental Politics* (Preprint) 22(2): 1–11. doi:10.1162/glep_a_00647.

Patterson, James J. 2023. "Backlash to Climate Policy." *Global Environmental Politics* 23(1): 68–90. doi:10.1162/glep_a_00684.

Pepermans, Yves, and Pieter Maeseele. 2016. "The Politicization of Climate Change: Problem or Solution?" *WIREs Climate Change* 7(4): 478–85. doi:10.1002/wcc.405.

Pettifor, Ann. 2020. *The Case for the Green New Deal.* London: Verso.

Pianta, Mario, and Matteo Lucchese. 2020. "Rethinking the European Green Deal: An Industrial Policy for a Just Transition in Europe." *Review of Radical Political Economics* 52(4): 633–41. doi:10.1177/0486613420938207.

Pickering, Jonathan, and John S. Dryzek. 2019. *The Politics of the Anthropocene.* New York, NY: Oxford University Press.

Rayner, Tim, and Andrew Jordan. 2013. "The European Union: The Polycentric Climate Policy Leader?" *Wiley Interdisciplinary Reviews: Climate Change* 4(2): 75–90. doi:10.1002/wcc.205.

Rietig, Katharina. 2019. "The Importance of Compatible Beliefs for Effective Climate Policy Integration." *Environmental Politics* 28(2): 228–47. doi:10.1080/09644016.2019.1549781.

Rietig, Katharina. 2021. "Accelerating Low Carbon Transitions via Budgetary Processes? EU Climate Governance in Times of Crisis." *Journal of European Public Policy* 28(7): 1018–37. doi:10.1080/13501763.2021.1918217.

Rietig, Katharina, and Claire Dupont. 2021. "Presidential Leadership Styles and Institutional Capacity for Climate Policy Integration in the European Commission." *Policy and Society* 40(1): 19–36. doi:10.1080/14494035.2021.1936913.

Rifkin, Jeremy. 2020. *The Green New Deal. Why the Fossil Fuel Civilization Will Collapse by 2028, and the Bold Economic Plan to Save Life on Earth.* New York: St Martin's.

Samper, Juan Antonio, Amanda Schockling, and Mine Islar. 2021. "Climate Politics in Green Deals: Exposing the Political Frontiers of the European Green Deal." *Politics and Governance* 9(2): 8–16. doi:10.17645/pag.v9i2.3853.

Schramm, Lucas, Ulrich Krotz, and Bruno De Witte. 2022. "Building 'Next Generation' After the Pandemic: The Implementation and Implications of the EU Covid Recovery Plan." *JCMS: Journal of Common Market Studies* 60(S1): 114–24. doi:10.1111/jcms.13375.

Selby, Jan. 2019. "The Trump Presidency, Climate Change, and the Prospect of a Disorderly Energy Transition." *Review of International Studies* 45(3): 471–90.

Siddi, Marco. 2021. "Coping with Turbulence: EU Negotiations on the 2030 and 2050 Climate Targets." *Politics and Governance* 9(3): 327–36. doi:10.17645/pag.v9i3.4267.

Skjaerseth, Jon Birger. 2021. "Towards a European Green Deal: The Evolution of EU Climate and Energy Policy Mixes." *International Environmental*

Agreements: Politics, Law and Economics 21(1): 25–41. doi:10.1007/ s10784-021-09529-4.

Skjaerseth, Jon Birger, Guri Bang, and Miranda A. Schreurs. 2013. "Explaining Growing Climate Policy Differences Between the European Union and the United States." *Global Environmental Politics* 13(4): 61–80. doi:10.1162/ GLEP_a_00198.

Steffen, Will. 2012. "A Truly Complex and Diabolical Problem." In *Oxford Handbook of Climate Change and Society*, eds. John S. Dryzek, Richard B. Norgaard, and David Schlosberg. Oxford: Oxford University Press, 21–37.

Stevenson, Hayley, and John S. Dryzek. 2014. *Democratizing Global Climate Governance*. Cambridge: Cambridge University Press.

Stokes, Leah. 2020. *Short Circuiting Policy: Interest Groups and the Battle over Clean Energy and Climate Policy in the American States*. Oxford: Oxford University Press.

Stokes, Leah C. 2016. "Electoral Backlash Against Climate Policy: A Natural Experiment on Retrospective Voting and Local Resistance to Public Policy." *American Journal of Political Science* 60(4): 958–74. doi:10.1111/ ajps.12220.

Swyngedouw, Erik. 2022. "The Unbearable Lightness of Climate Populism." *Environmental Politics* 31(5): 904–25. doi:10.1080/ 09644016.2022.2090636.

Tosun, Jale, and B. Guy Peters. 2021. "The Politics of Climate Change: Domestic and International Responses to a Global Challenge." *International Political Science Review* 42(1): 3–15. doi:10.1177/0192512120975659.

UNFCCC. 2021. "The United States of America. Nationally Determined Contribution. Reducing Greenhouse Gases in the United States: A 2030 Emissions Target." https://unfccc.int/sites/default/files/NDC/2022-06/Uni ted%20States%20NDC%20April%2021%202021%20Final.pdf.

Weible, Christopher M., and Paul A. Sabatier, eds. 2017. *Theories of the Policy Process*. New York: Westview.

Weible, Christopher M., and Samuel Workman, eds. 2022. *Methods of the Policy Process*. Abingdon: Routledge.

Wendler, Frank. 2019. "The European Parliament as an Arena and Agent in the Politics of Climate Change: Comparing the External and Internal Dimension." *Politics and Governance* 7(3): 327–38. doi:10.17645/pag.v7i3.2156.

Wendler, Frank. 2022a. *Framing Climate Change in the EU and US After the Paris Agreement*. Basingstoke [u.a.]: Palgrave Macmillan.

Wendler, Frank. 2022b. "The Politicization of Climate Change Governance: Building Blocks for a Theoretical Framework and Research Agenda." *Working Paper No. 6 – September 2022, Universität Hamburg*.

Wendler, Frank. 2023. "The European Green Deal Agenda After the Attack on Ukraine: Exogenous Shock Meets Policy-Making Stability." *Politics and Governance* Special Issue: Governing the EU Polycrisis: Institutional

Change After the Pandemic and the War in Ukraine 11(4). doi: 10.17645/pag.v11i4.7343.

White House. 2022. "Building a Better America. A Guidebook to the Bipartisan Infrastructure Law for State, Local, Tribal, and Territorial Governments, and Other Partners." www.whitehouse.gov/wp-content/uploads/2022/05/BUILDING-A-BETTER-AMERICA-V2.pdf.

White House. 2023. "Building a Clean Energy Economy: A Guidebook to the Inflation Reduction Act's Investment in Clean Energy and Climate Action; Version 2." www.whitehouse.gov/wp-content/uploads/2022/12/Inflation-Reduction-Act-Guidebook.pdf.

Willis, Rebecca. 2020. *Too Hot to Handle? The Democratic Challenge of Climate Change*. Bristol: Bristol University Press.

Workman, Samuel, Frank Baumgartner, and Bryan Jones. 2022. "The Code and Craft of Punctuated Equilibrium." In *Methods of the Policy Process*, eds. Christopher Weible and Samuel Workman. Abingdon: Routledge, 51–79.

Wurzel, Rüdiger, James Connelly, and Duncan Liefferink, eds. 2017. *The European Union in International Climate Change Politics: Still Taking a Lead?* London: Routledge, Taylor & Francis group.

Wurzel, Rüdiger, Mikael Skou Andersen, and Paul Tobin, eds. 2021. *Climate Governance Across the Globe. Pioneers, Leaders and Followers*. Abingdon, OX [u.a.]: Routledge.

Zhou, Jack. 2016. "Boomerangs versus Javelins: How Polarization Constrains Communication on Climate Change." *Environmental Politics* 25(5): 788–811. doi:10.1080/09644016.2016.1166602.

Zürn, Michael. 2019. "Politicization Compared: At National, European, and Global Levels." *Journal of European Public Policy* 26(7): 977–95. doi:10.1080/13501763.2019.1619188.

2   Theoretical framework

Stability and punctuation in policy-making on climate change

The framework of analysis presented here adopts concepts of PET as an approach that seeks to explain the interaction of two dynamics in the policy process: namely, incremental policy-making and stability of the institutional and political parameters of policy processes on the one hand, and bursts of far-reaching policy change marked by disruption and conflict in the surrounding political environment on the other (Baumgartner et al. 2009, 2018, Baumgartner & Jones 2015, Workman et al. 2022; cp. Wendler 2023). While key components of the present approach are borrowed from PET, the subsequent analysis does not attempt a full application of its ambition and rationales, particularly with regard to the longitudinal analysis of agenda-setting and policy processes over time; the goal followed here is rather to harness PET for evaluating the interaction between politics and policy-making on climate change as a response to exogenous shock and political contestation as observed in the present two cases of the EU and US.

Three defining ideas of PET recommend its application to the present cases compared to other theories of the policy process (cp. Weible & Sabatier 2017 Workman & Weible 2022). First, as an approach that focuses on policy change through dynamics of agenda-setting, it offers explanations for how the political definition of complex policy problems can shift boundaries of policy-making fields. Second, the role of policy beliefs is highlighted as a set of ideas that are key for understanding dynamics of the policy process and explaining policy change or continuity. Finally, a key component of PET that is relevant for the present analysis is the interaction between two different institutional levels: namely, acts of general agenda-setting at the level of macropolitical institutions, and processes of issue-specific

DOI: 10.4324/9781003452041-2

decision-making within policy subsystems specialized in a particular field of decision-making. All of these three features fit to the analysis of climate governance processes that expand through processes of agenda-setting in response to external shock as observed in the EU and US cases.

The main rationale for choosing this approach is to provide answers about how the linkage of new ideas and beliefs to climate action affects policy-making in a comparison between conditions of stability and change. By choosing a general theory of the policy process, our aim is also to go beyond the application of specific theories of European integration such as post-functionalism that also combine theoretical components of framing, venue selection and policy feedback (Hooghe & Marks 2009, 2019).

2.1 Policy images, venues and feedback in decision-making on climate change

Three key concepts of PET are discussed in this section as the foundation for our theoretical framework: namely, *policy images* as a term for the variable and potentially contested definition and evaluation of complex policy challenges as a political problem at the stage of agenda-setting; their effect on the involvement of *policy venues* as relevant sets of institutions and their interaction within the distinction of the macropolitical level and policy subsystems; and finally, *policy feedback* as a concept for evaluating decision-making results as a vector of policy change or continuity (Baumgartner et al. 2018, Workman et al. 2022). The presentation of these components in subsequent sections goes through the three steps of clarifying each concept at a general level, explaining its more specific significance in the context of climate politics, and then outlining its operationalization for the present analysis. A summary and visual overview of the theoretical framework follows before the conclusion of this chapter.

2.1.1 Policy images: defining political problems at the stage of agenda-setting

In its most succinct definition, a policy image defines "how a policy is understood and discussed" (Baumgartner & Jones 2015: 25, cp. Baumgartner et al. 2018: 62ff.). More specifically, the concept is used to identify what policy beliefs are proclaimed by political agents to define the scope and nature of a given policy problem at the stage of agenda-setting. Furthermore, it identifies what possible paths are proposed for solutions in terms of concrete approaches, measures and instruments. In this sense, policy images are defined as a "mixture of empirical information and emotive appeals" (ibid.: 26) and evaluated as a discursive concept primarily from surveys of political speech and policy documents. Especially when applied to complex issues that allow multiple perspectives on the perception of a given problem such as climate change, policy images define agenda-setting in two ways: first, by proposing an emotive understanding of the urgency, relevance and evaluative standards to be applied in implicit or explicit comparison to other problem definitions; and second, by offering empirical statements and causal assumptions for justifying the suitability, feasibility and effectiveness of proposed policy-making solutions. Similar to concepts of framing analysis (cp. Wendler 2022: 35ff.), the reconstruction of policy images is productive as an analytical concept if it can be used to highlight the selectiveness and possible contentiousness of a particular policy-making perspective based on distinctive policy beliefs.

From the theoretical perspective of PET, the evaluation of policy images assumes a dual relevance. First, in a longitudinal view, changing policy images are analyzed as part of broader shifts in political attention in political publics between competing topics as prompted by changes in media coverage or exogenous shocks, and identified through surveys of quantitative data over time. Second, in a shorter time frame, the emergence of new policy images within a specific set of policy process is analyzed with regard to its implications for subsequent decision-making; this arguably requires a more in-depth and primarily qualitative review of agendas that give rise to policy change. These two perspectives differ by considering policy images as a result of political attention cycles in the former, and as a cause for subsequent policy change in the latter variant. While both the longer-term mapping and case-specific assessment of policy images play a role in the present study, it is particularly the latter meaning of policy images as a cause for change in the policy process that assumes importance for the subsequent analysis.

Applied to climate change as a highly complex or even "super wicked" policy problem (Grasso & Markovitz 2015, Levin et al. 2012, Steffen 2012, Incoprera 2016), the concept of policy image is relevant to distinguish and systematize the emotive appeals and empirical statements that contribute to the discourse politics of its proposal and negotiation (Paterson 2021b). From the outset, climate change is a multi-faceted issue with strong emotive aspects of ecological deterioration, threats to humans or even catastrophic consequences for society. At the same time, its political treatment is approached from multiple angles that involve a range of empirical claims relating to the link between action against climate change and issues of economic growth, social justice and security (Dirikx & Gelders 2010, Hayes & Knox-Hayes 2014). In the context of the political developments considered here, the link between concepts of a green deal, responses to exogenous shock and evaluations of growth paradigms plays a particularly critical role for the construction of policy images (Domorenok & Graziano 2023, Hickel & Kallis 2020, Ossewaarde 2020). Applying the concept for the present analysis follows the rationale of evaluating continuity and change: to what degree policy images change through the onset of exogenous shocks and political conflict through various stages of agenda-setting. In terms of operationalization, the subsequent analysis therefore applies three indicators to evaluate vectors of change that affect policy images defining climate action in the EU and US:

(1) The first indicator covers the *scope* of climate change as an issue for policy-making: in this regard, we inquire what dynamics of change are observable in the range of policy-making issues that are brought in relation with the concept of reducing GHG emissions as the defining target of climate action. As discussed at the outset, the framing of climate policy can range from a broad understanding as a cross-governmental issue to more specifically defined perspectives that relate climate action to ideas of economic recovery, geopolitical security or technological competitiveness (Wendler 2022, McDonald 2021, Dalby & O'Lear 2016). Applied to the evaluation of emotive appeals and empirical information by macropolitical agents at the stage of agenda-setting, the key aspect is to what degree the scope of policy images changes through expansion or shift of primary fields of action. Applied to the two case studies, this step covers communication by the European Council and Commission in the EU, and by the White House and key departments of the executive branch in the US.

(2) The second indicator engages with the content of policy images by evaluating the *priority* of climate action targets: to what degree initiatives for moving towards decarbonization are foregrounded

in relevant statements, and what density related measures have across the entire range of actions and policies covered in green industrial policy agendas. In this sense, climate action can be proclaimed as a highly prioritized, primary target of policy-making that defines approaches to all involved fields of action through the priority of decarbonization. An observation especially in the context of the "green" industrial and recovery programs considered here, however, is that these targets are often considered more as a positive side effect of policies that are promoted in the pursuit of other, more traditionally socio-economic goals such as growth or competitiveness.

(3) The third indicator evaluates the *transformative appeal* inherent in policy images as a defining rationale of climate action: to what degree emotive appeals and empirical information call for change of extant policy-making related to climate change or, indeed, confirm ideas of stability and continuity. As in the previous two indicators, the key focus is on the evolution of policy images and to what extent they change towards more transformative or more stability-oriented appeals. For gauging the degree of stability or change proclaimed through policy images, a useful benchmark is the distinction between three levels of policy beliefs found across the theoretical literature on policy change: namely, between those relating to paradigms as core defining ideas of policy-making; those relating to programs as concepts describing the broader framework and general approach to the choice and application of policy instruments; and those specifying policy-related instruments as more specific aspects of particular components of action against climate change (Weible & Sabatier 2017 140f., Workman & Weible 2022: 109). Applying this threefold distinction, appeals for change projected through policy images can emerge at different levels: calls for paradigmatic change to society, such as those based on ideas of planetary survival or social justice (Pickering & Dryzek 2019; calls for programmatic change such as a shift of core policy goals towards a more geopolitical orientation; or finally, appeals relating to more policy-specific issues such as the setting of carbon prices or provision of incentives for investment in green technologies.

Combining the three indicators presented here – namely, the scope, priority and transformative appeal of policy beliefs about climate change – the focus of the present analysis is laid on the impulse for policy change that arises from a policy image: based on these criteria, these can range from highly salient calls for transformative change to more stability-oriented and less visible, even disperse, proposals for incremental and more stability-oriented policy-making.

2.1.2 Policy venues: decision-making between macropolitics and policy subsystems

How ideas embodied in policy images affect the choice and inter-action of institutional settings that become involved in a policy process is a question at the core of PET (Baumgartner et al. 2018: 59). In a general sense, policy venues are defined as "the institutional locations where authoritative decisions are made concerning a given issue" (Baumgartner & Jones 2015: 32). In our present cases, this applies primarily to legislative institutions such as the Council of the EU and the EP, and both chambers of the US Congress, and their specific procedures for legislative negotiation through committees and arrangements for inter-institutional negotiation. It also includes relevant action of executive bodies and independent agencies, as well as mechanisms of inter-departmental coordination and direction by central executive institutions. How policy images and venues interact is discussed in two ways by PET.

First, it is assumed that by recommending particular understandings of relevant portfolios, competence and expertise, policy images are tied to institutional organs and settings that are empowered when their respective field of competence is highlighted. For instance, agendas of climate action that are informed by policy images with a focus on technological, security, agricultural or economic aspects would refer to the expertise and responsibility of different legislative committees, agencies and executive departments (Baumgartner et al. 2018: 62ff.; Workman et al. 2022: 66). In this sense, it is suggested that a rapid shift of a policy image of climate action (e.g., towards a more clearly geopolitical approach) prompts an involvement of new venues in the policy process, and thereby disrupts extant policy monopolies established by longer-standing ownership of an issue by a committee or agency.

Second, how linkages between policy images and venues come about depends critically on the interaction between two different sets of institutions: on the one hand, those at the macropolitical level tasked with political leadership and agenda-setting primarily through executive agents and institutions; and on the other, the sets of institutions described as policy subsystems with specific mandates, competence and expertise in a particular field with often long-standing trajectories of decision-making in a given policy-making field (Cejudo & Trein 2023). PET emphasizes that both sets of institutions relate to observations of policy stability and change through the different forms of processing information for which they are suited: in this sense, macropolitical institutions operate through a serial treatment of topics and attention – one after the other, with rapid shifts following events of external shock – and thereby tend to provide impulses for change in dominant understandings of policy and their ranking as urgent or irrelevant. Policy subsystems, in turn, support a parallel processing of information – one next to another, with no direct interaction or competition – by allowing different aspects of broad political agendas to be compartmentalized (Baumgartner et al. 2018: 59f.). Within this logic, a widening or shift of policy images is accommodated by allocating different tasks to settings with specific expertise in their respective field, thereby stabilizing extant allocation of tasks and competence and even creating barriers to substantial policy change. A critical factor for policy and institutional change to occur, therefore, is the structure of interactions between both levels: to what degree macropolitical institutions interfere in the allocation of tasks or even proceed to create new institutions and policy subsystems.

For policy processes dealing with climate change, this theoretical discussion is intriguing at a time during which it receives an increased amount of attention at the macropolitical level in both the EU and US and has moved towards the top of the agendas of the political leaderships of both jurisdictions. As discussed before, however, the salience of climate change as an issue is also affected by a series of exogenous shocks that may have affected its priority as an agenda-setting issue. Unlike issues such as air pollution that have a clearly assigned status as a subset of environmental policy, climate action also involves a broad and variable range of different policy subsystems covering energy, trade, agriculture or budget issues in addition to more classic environmental issues such as land use or biodiversity. Especially in the cases covered here, the institutional boundaries of climate action are therefore more likely to be re-defined by shifts of agenda-setting at the macropolitical level than in other fields. As a consequence, new tasks of green investment are assigned to policy subsystems that are either newly established or were previously outside the perceived boundaries of climate governance. The discussion of how change in policy images affects institutional settings therefore points to an aspect of key relevance for the cases of climate governance considered here. The subsequent analysis again applies three indicators to evaluate the dynamics of stability and change at the level of policy venues:

(1) The first indicator gauges the *vertical intervention of* macropolitical agents into policy subsystems involved in climate governance processes: more specifically, to what extent initiatives by macropolitical agents are discernible that concern the allocation of tasks, the relative authority of climate- and environment-related agents in relation to other subsystems, or even the establishment of new institutional frameworks and procedures of governance. Here, interaction can range from direct, top-down direction and intervention to continuity of decision-making in subsystems without interference from the macropolitical level.

(2) A second indicator reviews the *horizontal friction* between involved policy subsystems and their relative degree of independence from one another (Baumgartner et al. 2018: 69, Workman et al. 2022: 69ff.). In this sense, aspects of agendas can be assigned to different specialized subsystems in a logic of parallel processing of related but separate components of a policy process; examples include separate decision-making on regulatory and expenditure-related aspects of a green recovery agenda or the assignment of different regulatory tasks covering agriculture, buildings and mobility to agencies and departments in their respective fields of expertise. By contrast, interaction and disruption between policy subsystems can occur through functional links such as the enforcement of cross-cutting regulatory standards, and particularly through linkages and package deals. In this regard, green recovery packages potentially create linkages of policy-making procedures to fields and decision-making previously considered external to climate action, such as budgetary negotiations or a combined negotiation of macroeconomic, healthcare and energy transition issues. Strong linkages between policy subsystems are therefore more likely to create political conflict and disruption, whereas independence of subsystems from one another is more conducive to a parallel processing of policy-making tasks.

(3) A third indicator evaluates the *external accountability* of policy-making as the emergence of links between policy subsystems and political publics through forms of political and electoral visibility and possible sanctioning. A stronger link is created when the assignment of policy tasks and decision-making between and in policy subsystems is subject to strong mechanisms of public and electoral scrutiny with a high public visibility and salience as a party political issue. Assignment of tasks to independent agencies, informal decision-making in settings with little visibility, and the absence of party political interaction by compound representation are considered as factors weakening this link. In particular, the legislative process of the EU seems to be unique in minimizing its external visibility and accountability through mechanisms of informal negotiation through trilogue (Bressanelli et al. 2016, Kreppel 2018). In comparison, legislative politics in the US is often more strongly exposed to political publics.

To summarize, the three indicators reviewed here – vertical and horizontal interaction of policy venues and their exposure to mechanisms of political accountability – are used to distinguish between two dynamics of policy-making: on the one hand, a disruptive top-down intervention of macropolitical agents into subsystem decision-making prompting punctuated rather than gradual change of venues and decision-making; and on the other, a more negotiated and informal assignment of policy-making tasks to related but separate policy subsystems tasked with a parallel processing of different policy-making items without direct mutual interference and with weaker external accountability.

2.1.3 Policy feedback: evaluating vectors for policy change

As the third analytical stage of our theoretical model, policy-making results are evaluated within the broad dualism of positive and negative feedback as discussed by PET (Baumgartner et al. 2018: 70ff., Workman et al. 2022: 68ff.). In its definition, the former is associated with disruptive changes of policy images and venues that effect far-reaching policy change; the latter is used as a concept for incremental and continuous policy-making within policy monopolies that remain resilient against external disruption or macropolitical intervention (Cejudo & Trein 2023, Stokes 2020). Negative feedback, in this sense, is associated with the concept of policy stability and the pursuit of a prescribed, longer-term policy-making trajectory towards established targets and following unchanged priorities.

Applied to the field of climate governance, the concept of policy change can be a confusing one, as the steady pursuit of decarbonization targets along a predefined path is not conceivable without the successful initiation of far-reaching change in a wide variety of societal and economic fields (Latour 2018, Pickering & Dryzek 2019). So does the steady realization of a zero-carbon agenda over time signify policy change in this respect or its presumable contrary, policy stability? Another difficulty stems from the fact that the simple observation of policy change can stand for either increased ambitions of climate policy or its repeal, making a clearer distinction of the direction of policy change necessary.

From this point of departure, we posit that rather than the occurrence of policy change as such, its speed and direction in relation to previous stages of policy development are critical for evaluating the impact of decision-making on policy against climate change. In this sense, policy stability (or "negative feedback" in PET language) stands for the pursuit of policies that enact societal changes to achieve decarbonization at about the same speed and with a steady approach and rationale. By contrast, a departure from established policy trajectories is indicative of "positive feedback" as a term for more disruptive policy change, and in principle covers both advances towards more stringent climate action and its repeal. Based on this understanding, the evaluation of policy-making results as positive or negative feedback is again based on three related indicators:

(1) The first indicator evaluates the effect of decision-making on the *speed* of policy development towards the overall target of decarbonization, measured in its observed or expected contribution to lowering GHG emissions. While the distinction between positive and negative feedback becomes a gradual one through the quantification of reductions in GHG emissions, disruptive policy change is associated with a significant adjustment of longer-term targets to reduce emissions, whereas policy stability is found in decisions to reaffirm, support or enact these targets. Set in context with previous policy-making stages, this criterion creates a better distinction between progressive policies that evolve in a dynamic of policy stability – namely, through continuous policy

adjustment to achieve set goals of decarbonization – from disruptive change that affects the ambition of climate policy as such through decision-making about acceleration or slowdown.

(2) A second indicator gauges the *stringency* applied to achieve a reduction of GHG emissions, in terms of applied policy instruments and their combination (Oberthür & von Homeyer 2023). A key aspect in this regard is the mix of policy instruments applied to achieve targets of climate action: to what degree a reduction of carbon emissions is explicitly prescribed through binding regulation, or instead limited to incentives and guidance that encourage but do not enforce policy change. As a benchmark for evaluating the stringency of policies, we apply a three-step classification of instruments in descending order from prescriptive (or "harder") to more flexible and open-ended (or "softer") variants: the setting of restrictive regulatory standards requiring the achievement of an explicit target or standard of emission reductions; the adoption of incentive-based policies through direct or indirect financial subsidies and facilitation of regulatory standards to encourage investment and consumer behavior in favor of emission reductions; and finally, advisory or coordinative governance mechanisms encouraging decision-making in favor of decarbonization targets. In this evaluation, policy change increases (or weakens) the stringency of climate action by adding or reinforcing relatively more binding types of instruments in comparison to more voluntary ones. Policy stability, by contrast, is identified through a steady combination of instruments with unchanged stringency over time.

(3) Finally, as a reflection of a key argument of the policy stability literature – namely, the ambition of adopting climate policies that are as close to irreversible as possible – the last indicator evaluates the *durability* of policy change in terms of its temporal duration and vulnerability to modification or repeal (cp. Carlson & Burtraw 2019, Jordan & Matt 2014, Jordan & Moore 2020). Aspects to be considered in this regard concern the time frame of subsidies, tax breaks and other forms of public financing; the strength of legally binding commitments to achieve targets of decarbonization; and the probability of legal challenges to or political repeal of adopted policies. This indicator therefore adds an important aspect to the discussion of policy change and stability: namely, whether disruptive breakthroughs to substantial

policy change can be expected to last for longer periods of time and solidify into policy-making stability, or whether they are followed by periods of volatility.

By specifying the dualism of positive and negative feedback through the three indicators presented here, the difficulties associated with evaluating policy change in the specific field of climate governance are resolved in two ways. First, we create a distinction between primary policy change affecting the establishment and consolidation of longer-term trajectories of climate action, and secondary decision-making related to the adjustment of social and economic policies required to enact these trajectories. Second, the error of conflating policy change in the two opposed directions of more ambitious climate action and its repeal is avoided particularly by evaluating the speed and stringency of climate policies.

To summarize, the theoretical approach presented here evaluates how ideas (*policy images*), institutions (*policy venues*) and subsequent results of decision-making (*policy feedback*) interact to explain the dynamics of policy stability and change. As explained above, two mechanisms envisaged by PET are particularly decisive for the occurrence of these two dynamics: first, the emergence of new policy images requiring an adjustment of extant policy processes; and second, an intervention of macropolitical agents into policy subsystems that enacts related shifts and results in policy change. As Baumgartner and Jones succinctly put it: "In summary, subsystem politics is the politics of equilibrium, [...], a widely supported policy image, and negative feedback. Macropolitics is the politics of punctuation, [...] large-scale change, competing policy images, [...] and positive feedback" (Baumgartner et al. 2018: 63). The theoretical framework is summarized in Figure 2.1.

The theoretical framework presented here informs the research question described at the outset in two ways. First, the model can be applied to specify vectors of relative policy-making stability versus those that are related to political conflict and disruption. While it identifies precise indicators for evaluating change or continuity through stages of the policy process dealing with agenda-setting, policy formulation and decision-making as well as subsequent results, the embedding of this framework in PET also indicates ways through which the three stages are connected. In this sense, we adopt its core assumption that rapid shifts of policy images will have a disruptive impact on

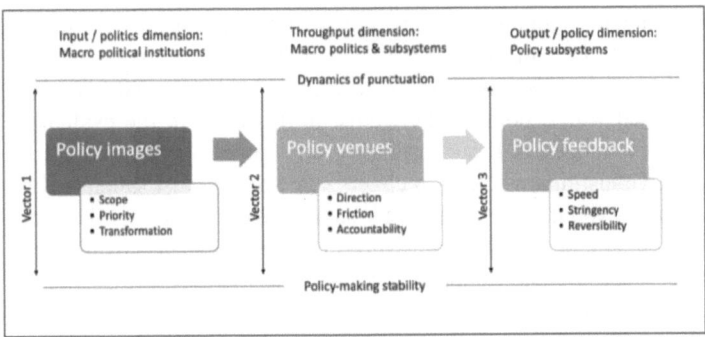

Figure 2.1 Overview of the theoretical framework

policy venues and result in breakthroughs to positive feedback, in contrast to trajectories characterized by stability and continuity of policy-making. Pressures for change arising from new policy images, however, are mediated through the resilience of policy subsystems and their interaction with macropolitical institutions as described.

Second, by highlighting contrasts between stability and change, the framework is suited for comparing climate policy-making in the EU and US as two likely cases of these two contrasting dynamics of policy development. A common point of reference for both is the expansion of climate agendas through the adoption of climate action as a cross-cutting issue and subsequent green industrial policies. As the case studies will show, however, this expansion evolves through different mechanisms in the two cases of the EU and US: namely, one marked by incremental additions to a firmly established climate agenda and relatively stable core of decision-making venues and policy subsystem in the EU; and by contrast, in a more contested and disruptive sequence of decisions creating a new governance frame-work and breakthrough for climate action in the US.

2.2 Data and methods

As the focus of the present analysis is laid on stages of the policy process from agenda-setting to decision-making, its empirical material primarily consists of policy documents issued by executive institutions and negotiated by legislative agents through the stages of this process.

Fitting to the analytical focus of the theoretical model, the selection of documents includes documentation of broad political guidelines proclaimed at the macropolitical level by central executives, but then proceeds to coverage of more issue-specific programs and initiatives negotiated within policy subsystems. In this context, the evaluation of policy images is based primarily on written agendas issued by executive institutions but also covers some of the most salient public statements about these agendas by senior executive actors. For more issue-specific decision-making, the subsequent analysis covers legislative proposals and adopted texts but also makes use of officially issued summaries of legislative negotiations by services such as the Congressional Research Service and European Parliament Research Service. The precise selection of documents is presented in the respective case study chapters on the EU and US and listed in full detail in the Appendix.

Concerning policy results, a particular challenge of the present analysis is that implementation of major legislative acts such as the IRA in the US is still ongoing, and actual effects in terms of decarbonization and other aspects are only gradually emerging. In this regard, the analysis evaluates reports and prognoses about the impact of green recovery programs issued by executive departments and independent think tanks with recognized expertise in the modeling of emission reductions and other policy effects.

Concerning the choice of methods, the subsequent analysis combines elements of quantitative content analysis for surveys of large bodies of text with more in-depth qualitative analysis of those documents and decision-making procedures that appear key for defining policy images and shaping decisions. Complete overviews of coding categories and keywords as well as details concerning the coding software and methodology used for quantitative content analysis are included in the Appendix.

2.3 Summary: policy stability and punctuation as vectors of climate governance

The point of departure for the comparison of climate governance in the EU and US in this volume is to evaluate dynamics of change: while climate policy has risen to the top of political agendas and become more visible and contentious as a political topic, its definition and boundaries as a field of policy-making have been re-defined as a response to a chain of exogenous shocks and as a result of political contestation. In this setting, it seems relevant to investigate the interaction of vectors of stability and disruption in evolving climate governance frameworks and their effect on policy-making results.

The theoretical framework presented in this chapter harnesses PET as a well-established theory of the policy process to trace vectors of stability and punctuation in the evolution of agendas, decision-making frameworks and policy results. The application of the theoretical framework in the subsequent analysis is presented in two steps. In the following two chapters we explore the evolution of climate change governance frameworks in the two cases of the EU and US (Chapters 3 & 4). A comparative evaluation and discussion of main findings concerning the research question is presented in the conclusion (Chapter 5).

References

Baumgartner, Frank, and Bryan Jones. 2015. *The Politics of Information. Problem Definition and the Course of Public Policy in America.* Chicago: University of Chicago Press.

Baumgartner, Frank, Bryan Jones, and Peter B. Mortensen. 2018. "Punctuated Equilibrium Theory: Explaining Stability and Change in Public Policymaking." In *Theories of the Policy Process*, eds. Christopher M. Weible and Paul A. Sabatier. New York, NY: Westview Press, 55–102.

Baumgartner, Frank, Christian Breunig, Green Pedersen, Bryan Jones, Peter B. Mortensen, Michiel Nuytemans, and Stefaan Walgrave. 2009. "Punctuated Equilibrium in Comparative Perspective." *American Journal of Political Science* 53(3): 603–20.

Bressanelli, Edoardo, Christel Koop, and Christine Reh. 2016. "The Impact of Informalisation: Early Agreements and Voting Cohesion in the European Parliament." *European Union Politics* 17(1): 91–113. doi:10.1177/1465116515608704.

Carlson, Ann, and Dallas Burtraw, eds. 2019. *Lessons from the Clean Air Act. Building Durability and Adaptability into U.S. Climate and Energy Policy.* Cambridge: Cambridge University Press.

Cejudo, Guillermo M., and Philipp Trein. 2023. "Pathways to Policy Integration: A Subsystem Approach." *Policy Sciences* 56(1): 9–27. doi:10.1007/s11077-022-09483-1.

Dalby, Simon, and Shannon O'Lear. 2016. *Reframing Climate Change: Constructing Ecological Geopolitics.* Abingdon, OX; New York, NY: Routledge.

Dirikx, Astrid, and Dave Gelders. 2010. "To Frame Is to Explain: A Deductive Frame-Analysis of Dutch and French Climate Change Coverage During the Annual UN Conferences of the Parties." *Public Understanding of Science* 19(6): 732–42. doi:10.1177/0963662509352044.

Domorenok, Ekaterina, and Paolo Graziano. 2023. "Understanding the European Green Deal: A Narrative Policy Framework Approach." *European Policy Analysis* 9(1): 9–29. doi:10.1002/epa2.1168.

Grasso, Marco, and Ezra M. Markowitz. 2015. "The Moral Complexity of Climate Change and the Need for a Multidisciplinary Perspective on Climate Ethics." *Climatic Change* 130(3): 327–34. doi:10.1007/s10584-014-1323-9.

Hayes, Jarrod, and Janelle Knox-Hayes. 2014. "Security in Climate Change Discourse: Analyzing the Divergence Between US and EU Approaches to Policy." *Global Environmental Politics* 14(2): 82–101. doi:10.1162/GLEP_a_00230.

Hickel, Jason, and Giorgos Kallis. 2020. "Is Green Growth Possible?" *New Political Economy* 25(4): 469–86. doi:10.1080/13563467.2019.1598964.

Hooghe, Liesbet, and Gary Marks. 2009. "A Postfunctionalist Theory of European Integration: From Permissive Consensus to Constraining Dissensus." *British Journal of Political Science* 39(1): 1–23. doi:10.1017/S0007123408000409.

Hooghe, Liesbet, and Gary Marks. 2019. "Grand Theories of European Integration in the Twenty-First Century." *Journal of European Public Policy* 26(8): 1113–33. doi:10.1080/13501763.2019.1569711.

Incropera, Frank. 2016. *Climate Change: A Wicked Problem. Complexity and Uncertainty at the Intersection of Science, Economics, Politics, and Human Behavior*. Cambridge: Cambridge University Press.

Jordan, Andrew, and Brendan Moore. 2020. *Durable by Design? Policy Feedback in a Changing Climate*. Cambridge: Cambridge University Press.

Jordan, Andrew, and Elah Matt. 2014. "Designing Policies That Intentionally Stick: Policy Feedback in a Changing Climate." *Policy Sciences* 47(3): 227–47.

Kreppel, Amie. 2018. "Bicameralism and the Balance of Power in EU Legislative Politics." *The Journal of Legislative Studies* 24(1): 11–33. doi:10.1080/13572334.2018.1444623.

Latour, Bruno. 2018. *Down to Earth. Politics in the New Climatic Regime*. Cambridge: Polity Press.

Levin, Kelly, Benjamin Cashore, Steven Bernstein, and Graeme Auld. 2012. "Overcoming the Tragedy of Super Wicked Problems: Constraining Our Future Selves to Ameliorate Global Climate Change." *Policy Sciences* 45(2): 123–52. doi:10.1007/s11077-012-9151-0.

McDonald, Matt. 2021. *Ecological Security: Climate Change and the Construction of Security*. Cambridge, New York: Cambridge University Press.

Oberthür, Sebastian, and Ingmar von Homeyer. 2023. "From Emissions Trading to the European Green Deal: The Evolution of the Climate Policy Mix and Climate Policy Integration in the EU." *Journal of European Public Policy* 30(3): 445–68. doi:10.1080/13501763.2022.2120528.

Ossewaarde, Marinus, and Roshnee Ossewaarde-Lowtoo. 2020. "The EU's Green Deal: A Third Alternative to Green Growth and Degrowth?" *Sustainability* 12(23): 9825. doi:10.3390/su12239825.

Paterson, Matthew. 2021a. *In Search of Climate Politics*. Cambridge: Cambridge University Press.

Paterson, Matthew. 2021b. "'The End of the Fossil Fuel Age'? Discourse Politics and Climate Change Political Economy." *New Political Economy* 26(6): 923–36. doi:10.1080/13563467.2020.1810218.

Pickering, Jonathan, and John S. Dryzek. 2019. *The Politics of the Anthropocene*. New York, NY: Oxford University Press.

Steffen, Will. 2012. "A Truly Complex and Diabolical Problem." In *Oxford Handbook of Climate Change and Society*, eds. John S. Dryzek, Richard B. Norgaard, and David Schlosberg. Oxford: Oxford University Press, 21–37.

Stokes, Leah. 2020. *Short Circuiting Policy: Interest Groups and the Battle over Clean Energy and Climate Policy in the American States*. Oxford: Oxford University Press.

Weible, Christopher M., and Paul A. Sabatier, eds. 2017. *Theories of the Policy Process*. New York: Westview.

Weible, Christopher M., and Samuel Workman, eds. 2022. *Methods of the Policy Process*. Abingdon: Routledge.

Wendler, Frank. 2022. *Framing Climate Change in the EU and US After the Paris Agreement*. Basingstoke [u.a.]: Palgrave Macmillan.

Wendler, Frank. 2023. *The European Green Deal Agenda after the Attack on Ukraine: Exogenous Shock Meets Policy-Making Stability*. Politics and Governance Special Issue: Governing the EU Polycrisis: Institutional Change After the Pandemic and the War in Ukraine 11(4). doi: https://doi.org/10.17645/pag.v11i4.7343.

Workman, Samuel, Frank Baumgartner, and Bryan Jones. 2022. "The Code and Craft of Punctuated Equilibrium." In *Methods of the Policy Process*, eds. Christopher Weible and Samuel Workman. Abingdon: Routledge, 51–79.

3 European Union

Adjustment and expansion of the European Green Deal agenda

Since its proclamation in 2019, the EGD is the defining concept of EU climate action (COM 2019, Oberthür & Homeyer 2023, Skjaerseth 2021, Bongardt & Torres 2022, Eckert 2021, Samper et al. 2021, Siddi 2020, Wendler 2022, Schunz 2022, Ossewaarde et al. 2020, Delbeke & Vis 2019). Within this period, it has almost continuously evolved in a context of exogenous shocks that required its adaptation (Dupont & Torney 2021, von Homeyer et al. 2021, 2022, Anghel & Jones 2023, Kreienkamp et al. 2022). Originally, it was announced as one of the six headline ambitions of the incoming Commission under the leadership of President von der Leyen and proposed as part of its pledge to pursue a more clearly geopolitical ambition. Particularly concerning this latter aspect, it is widely perceived as a significant new step in the development of EU climate policy beyond the realm of environmental policy (Lenschow & Gravey 2021) and a step towards further expansion beyond the previous close linkage between climate and energy policy (Skjaerseth et al. 2017, Dupont 2016, Dupont & Oberthür 2012, Delbeke 2015, Boasson 2013).

Almost immediately, its launch was followed by the onset of the Covid-19 pandemic and subsequent debate about economic stimulus packages and policies aiming at dealing with the social impacts of lockdown measures. The main result was the inclusion of "green" criteria as a priority of EU recovery spending programs adopted under the framework of the NGEU program (Schramm et al. 2022, Buti & Fabbrini 2023, COM 2020, EPRS 2022, 2023e, Ekerbout et al. 2020, Ryner 2023). The EU climate agenda encountered its next challenge through the Russian war of aggression against Ukraine in February 2022. Among the implications of this event for

DOI: 10.4324/9781003452041-3

the EU climate agenda are a shift of political debate and attention to security issues, but also more directly economic effects on energy supplies and prices potentially affecting efforts to achieve a clean energy transition in the EU. In this regard, the REPEU program, adopted in May 2022 as the main political response by the EU to the war, has sought to create synergy between the targets of independence from Russian fossil fuel supplies and political efforts to achieve decarbonization (COM 2022a, 2022b, Wendler 2023, Vezzoni 2023, EPRS 2023a, Bonciu 2022).

Finally, the passage of legislation for green investment in the US, covered in more detail in the next chapter, has created the most recent challenge to the EGD agenda. In general terms, the IRA is considered welcome from an EU standpoint as a breakthrough for more stringent climate action by the US. However, it is also perceived as a challenge for the transatlantic relationship because of its protectionist elements and as a source of attracting investment in "green" industries and technologies away from the EU to the US (EPRS 2023c). Responses beyond the creation of an EU–US task force (COM 2023a, EPRS 2023d) have been proposed through the GDIP launched in early 2023 and are currently debated with regard to the Net-Zero Industrial Plan and related initiatives (COM 2023b,c, EPRS 2023f,g). This short survey demonstrates how the EGD agenda has been revised and adapted as a response to a sequence of exogenous shocks and challenges since its launch. The adaptation to these challenges is reflected in an extended literature on EU climate governance in a setting of turbulence (Siddi 2021) and crisis (von Homeyer et al. 2021, 2022, Siddi 2023, Burns 2019), and more widely embedded in the extended literature of the EU's observed disequilibrium (Hodson & Puetter 2019) and "polycrisis" (Zeitlin et al. 2019, Bressanelli & Natali 2023).

The process through which the EGD agenda and policy process have been adapted to this sequence of events is the subject of this chapter. A recurring theme of this reconstruction is the expansion of the governance process associated with the EGD. This dynamic can be traced from the foundation of the EGD in a regulatory framework establishing rules promoting reductions of carbon emissions to the addition of a green investment mechanism envisaged as part of the recovery mechanism established in the pandemic. It has been further modified towards a more explicitly geopolitical approach in the recent response to the war in Ukraine promoted through REPEU. Tracing this evolution, a focus of the chapter is to demonstrate how the governance framework established for enacting the EGD has unfolded while maintaining mechanisms associated with elements of policy-making stability. A related aim is to identify and evaluate the factors that create the relatively high degree of resilience of the EGD governance framework against events of exogenous shock.

Based on the theoretical framework presented above, the subsequent analysis applies the three related concepts of policy image, venues and feedback to trace this evolution. This proceeds in three steps: first, evaluating how the initial agenda and policy priorities of the EGD were adapted to the challenges posed by the pandemic and war in Ukraine; second, evaluating how the governance framework and involved policy subsystems were adjusted to deal with additional challenges resulting from these events; and finally, considering how policy change can be evaluated with regard to the initial targets of the EGD, and whether it involves observable policy shifts in the sense of a re-direction or reversal of policy-making. These three components are first applied in the next three sections (3.1–3.3), before main findings from this reconstruction are summarized with regard to the three core theoretical concepts and related to one another (3.4). The concluding part summarizes the main findings (3.5).

3.1 Policy images: adapting the European Green Deal agenda to new challenges

The EGD agenda is a program that promotes a strong policy image of emotive appeals and empirical statements, not least through the implicit historical reference of its title and high salience as a defining political program of the Commission. A first step to approach its adaptation to exogenous events is to map the political environment shaping the evolution of the EGD agenda. According to PET, an important driver of changing policy images are shifts of political attention and resulting changes in the prioritization of policy issues at the macropolitical level. The present analysis therefore starts with a survey of how the general salience of climate-related issues has developed at the level of EU agenda-setting in the period between the proclamation of the EGD in 2019 and the present stage. For this purpose, we reconstruct the evolution of agenda-setting by two EU institutions: first, the European Council as the highest executive institution of the EU, in charge of defining its overall political priorities and agenda (Rosamond & Dupont 2021); and second, the Commission as the primary agent for defining the mid-term agenda following from the guidance of the European Council, and for specifying policy programs and legislative proposals.

The subsequent survey traces the thematic salience of different topics at the level of EU macropolitical institutions since the launch of the EGD agenda in December 2019. Based on a dictionary-based automated coding of documents, it compares the relative frequency of keywords associated with five core topics related to the EGD agenda – namely, climate change, energy, economic development, recovery and security – at three macropolitical levels associated with agenda-setting in the EU: first, the conclusions adopted by the European Council as a result of each of its meetings; second, thematic references in the State of the European Union addresses[1] delivered by President von der Leyen on an annual basis to the EP as an expression of key political priorities adopted by the Commission for the nearer-term future; and finally, thematic emphases in the Annual Work Programs (AWPs) of the Commission as a link between the core priorities expressed at the previous two levels and the proposal of more specific programs and actions in particular fields of policy-making. In this context, it is worth mentioning that each of the AWPs is structured around the six headline ambitions proclaimed in 2019 by the incoming Commission, the first of which is the EGD in every version of the AWP from 2019 to 2023.

Set in comparison with one another, the three levels of agenda-setting compared here represent layers that reach from the definition of core priorities at the highest macropolitical level to the meso-level proposal of more specific policy-making actions. The same dictionary of thematic keywords was applied across these levels; therefore, direct comparison of thematic emphases is possible both across time and between the levels of agenda-setting. The results are shown in Figures 3.1, 3.2 and 3.3 below.

The survey demonstrates drastic changes in the amount of political attention paid to the topics compared, particularly within the conclusions of the European Council: at this level, the evolution of topics is marked first by a shift to questions of recovery after the onset of the Covid pandemic in 2020; and even more clearly, a shift to security-related issues after the Russian invasion of Ukraine in 2022 for it to become the most salient topic in this period. In comparison, it is intriguing how these shifts are observable but less strongly expressed in the agenda-setting

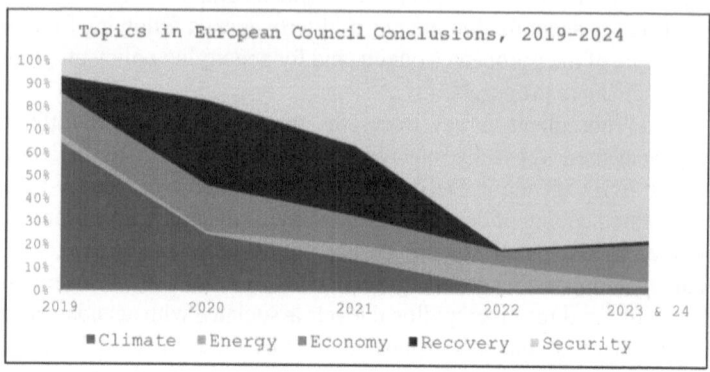

Figure 3.1 Topics in European Council conclusions, 2019–24

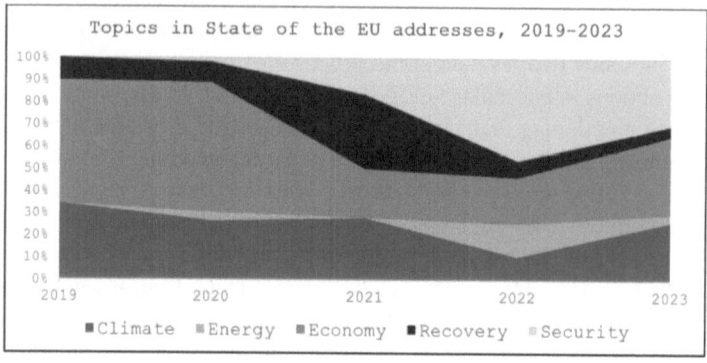

Figure 3.2 Topics in State of the EU addresses, 2019–23

of the Commission, at the level of both the State of the European Union addresses to the EP and particularly the AWPs. At both levels, an increased emphasis of references to recovery and security is not expressed to the extent observed at the level of the European Council and does not result in the removal of climate change as a salient topic. This indicates that even against the background of a more volatile political environment, the Commission has maintained a relatively strong degree of continuity in its agenda-setting and maintained a focus on questions of climate change, particularly in its AWPs.

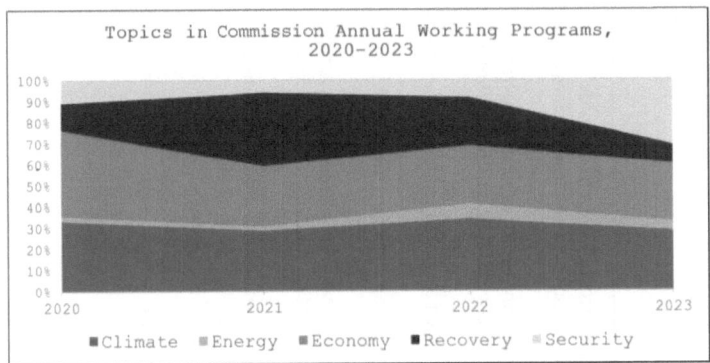

Figure 3.3 Topics in Commission Annual Work Programs, 2020–3

The temporal perspective on agenda-setting processes presented here establishes the background for a more in-depth, qualitative reconstruction of how policy images defining the EGD agenda have developed as a response to changes in its external political environment. In order to evaluate the evolution of policy images during the evolution of the EGD, a quantitative approach offers only limited opportunities for detecting argumentative linkages, particularly those associated with identification of emotive appeals and related empirical information. Combined with this methodical shift, our analytical perspective proceeds from exploring conditions of change of policy images to their case-specific reconstruction and subsequent impact on the policy process, as discussed in the theoretical chapter. As described above, four stages of agenda-setting can be distinguished in the evolution of the EGD. Based on the analysis of Commission documents and their review in the literature, the policy image of EU climate action has evolved through the following stages associated with the four sets of policy programs discussed above (summarized in Table 3.1):

(1) Launch of the EGD agenda: the policy image of the EGD essentially defines a new growth strategy for the EU, as reflected in the research literature on the concept of green growth and its critical evaluation (Eckert & Kovaleska 2021, Machin 2019, Domorenok & Graziano 2023, Ossewaarde 2020). As key emotive appeals, two core ideas stand out: on the one hand, the definition of

Table 3.1 Policy images of EU climate action through four stages of the European Green Deal agenda

Program	Policy image: characterization	Emotive appeals	Empirical information
EGD (2019)	EGD as Europe's new growth strategy	- Climate change as a generation's defining task - Transition to carbon neutrality will be fair and inclusive	- Economic growth can be decoupled from resource use - Two essential components are clean energy transition and circular economy - Transition presents opportunities for jobs and growth - Synergy can be created between economic environmental and social objectives - EU has capacity to proceed as global leader independently of other agents
NGEU (2020)	Green investment as sustainable response to the crisis	- Unprecedented crisis requires response defined by solidarity - Investment in next generation: long-term time horizon essential for action	- Drastic but asymmetrical impact of Covid-19 crisis on economy and society - New approach of target-oriented investment and spending - EGD as the first of three policy fundamentals for crisis response
REPEU (2022)	Clean energy transition strengthens EU geopolitical security	- Military aggression requires the end of dependence on Russian fossil fuels - Avoid unilateralism: EU must act in solidarity	- Continuation and reinforcement of energy transition and efficiency helps achieve fossil fuel independence - Diversification of fossil fuel supplies as temporary crisis management - Investment in infrastructure must be reinforced ("smart investment")

| GDIP (2023) | Adapting the EGD to an adverse global environment | - Present decade is decisive for creating net-zero industrial age of the future
- EU must act to combat effects of unfair subsidies in other world regions | - EU can be a leading player in net-zero industries of the future
- EGD requires massive increase in development and production of net-zero products
- Enactment of GDIP based on four pillars (regulation, funding, skills and trade)
- Enactment of GDIP through Net-Zero Industry Act focused on four pillars
- Creation of hydrogen bank and launch of competitive bid for hydrogen production
- REPEU explicitly included as strategy to support investment in net-zero industries |

Sources: COM 2019, 2020, 2021, 2022a, 2022b, 2023a; EPRS 2022, 2023c, 2023g; Machin 2019, Ossewaarde et al. 2020; Eckert 2021, Wendler 2022, 2023; Homeyer et al. 2021, Oberthür & Homeyer 2023, Vezzoni 2023.

climate change as a defining generational task requiring deeply transformative policies and a cross-cutting approach to promote climate action; and on the other, the appeal of designing this transformation in a way that preserves and strengthens social cohesion and fairness as two defining values of the European Social Model. The most central statements providing empirical information about approaches and causalities that follow from this approach include, most importantly, that economic growth can be made sustainable and decoupled from resource and ecological depletion when based on a clean energy transition, the creation of more resource-efficient infrastructure, and the establishment of a circular economy. Approached in this way, the EGD agenda is presented as a set of opportunities that creates synergy between key social, economic and environmental objectives rather than pitting them against one another. With regard to the international context, the EU is presented as a global leader of climate action that promotes cooperation but is ultimately able to act on its own to achieve its zero-carbon target, including through the use of a carbon border adjustment mechanism (CBAM).

(2) Green recovery in response to the Covid pandemic: while addressing the immediate economic and social impacts of the Covid-19 crisis, the main idea promoted by the Commission's key communication on its launch of NGEU is the adoption of a long-term perspective for designing the crisis response, as most centrally reflected in the reference to future generation in the program's title (Ryner 2023, Schramm 2022, Ekerbout et al. 2020, EPRS 2022). At the level of emotive appeals, the idea of solidarity between Member States is particularly emphasized to deal with the dramatic but highly asymmetrical impacts of the health crisis and subsequent economic downturn. At the level of empirical statements, the EGD is referenced as the first of three policy fundamentals next to the digital and socially fair transition, emphasizing the continuity of the Commission's agenda for dealing with the crisis. Aside from the discussion of financial instruments and governance mechanisms, the launch of NGEU is remarkable for the degree to which it maintains and reaffirms the defining concepts of the EGD agenda; its only major innovation is the adoption of a broader range of policy instruments including the mobilization of funds, and the closer policy integration with other investment strategies, particularly those in digital infrastructure.

(3) The REPEU agenda in response to the Russian attack on Ukraine: an immediate consequence of the Russian attack on Ukraine is an increased salience of security as a general policy-making principle, including its implications for energy supplies and prices in European countries. While this impact and the shift of attention to spending on military procurement could be seen as a challenge to efforts to achieve a transition to carbon neutrality, the REPEU agenda proclaims a renewed commitment to decarbonization as a project that creates synergy between ecological sustainability and energy independence and security (Wendler 2023, Vezzoni 2023, Bonciu 2022). Based on this understanding, the urge to achieve independence from Russian supplies of fossil fuels particularly stands out as the strongest emotive appeal in the REPEU communication; a second notable aspect is the appeal to Member States to act collectively and in solidarity with one another rather than choosing unilateral options as the only viable path to respond to the Russian aggression. Beyond this layer of emotive appeals and commitments, the most notable empirical component of the REPEU agenda is the proposition of a triangle consisting of the three related targets of energy efficiency, a clean energy transition and diversification of energy supplies as the appropriate policy tools to meet the proposed challenges. Notably, the first two components of this three-pronged strategy are defining elements of the EGD agenda while only the third – the ambition to diversify energy supplies, including the option to expand infrastructure for liquefied natural gas (LNG) storage and supply – creates a competing policy-making target. In addition to these three policy-making priorities, the commitment to the concept of "smart investment" in zero-carbon infrastructure creates a connection to the green recovery plans launched through the previous NGEU agenda. In this sense, the policy image proclaimed in response to the Russian attack adopts a more strategic and geopolitical approach to EU energy and climate governance while reconfirming rather than questioning its defining components (Wendler 2023, Siddi & Prandin 2023); the only potential deviation from key targets of the EGD agenda is the discussion of ad hoc measures taken by Member States to meet immediate or expected shortages of supply and invest in a diversification of fossil fuel supplies.

(4) Response to geopolitical and economic environment: the GDIP builds on the more geopolitical outlook of the previous REPEU

plan but applies it to the promotion of European net-zero industries and products in a context of more competitive and volatile relations on a global scale. The key emotive appeal is the description of the current decade as the decisive opportunity to create and design future markets for relevant industries, and to ensure the EU's ability to improve its competitiveness and reduce independence from external suppliers. In its global outlook, the program deviates from the initial EGD agenda by more decisively highlighting the need for the EU to act against unfair subsidies and trade practices and respond to the more adverse international environment following Russia's attack on Ukraine. The core policy idea to meet this challenge is defined as an "open but assertive approach" of the EU towards its global environment (COM 2023). Concerning its empirical statements, the GDIA builds on previously proposed programs such as faster permitting and investment in key technologies but adds further initiatives, particularly in the fields of skills creation and trade, including supplies of critical raw materials. More specifically, the Net-Zero Industry Act is proposed as a bundle of measures to promote EU competitiveness in critical technologies through better regulation, easier access to relevant funding (including the Temporary Crisis Framework for more flexible state aid rules for Member States), skills promotion, and re-orientation of trade agendas.

Based on this review, two major dynamics can be identified that have shaped the evolution of the policy image of the EGD agenda as the defining concept of EU climate action. The first is an expansion in the scope and ambition of this agenda particularly with regard to its envisaged range of policy instruments: most importantly, the initial emphasis on regulation to create a framework for the transition to carbon neutrality has been complemented by an additional governance framework to mobilize resources for a new version of supply-side green industrial policy and investment in decarbonized infrastructure. The second major dynamic of change is a more decisive turn towards a strategic and geopolitical approach of EU climate action. Going beyond the consideration of a CBAM in the initial EGD agenda, recent stages consider questions of external energy security and independence, and seek to secure the EU's position in competitive markets that are relevant for the green transition and for critical supply chains. A defining aspect of this evolution of agenda-setting policy programs, however, is that key targets and concepts of the original EGD agenda – in particular, its commitment to achieve carbon neutrality and its defining foundation in a re-definition of the EU's growth and social model – are retained and reconfirmed throughout the sequence of policy documents covered here. In this sense, the exogenous shocks identified at the outset have added to and complemented rather than re-defined the core goals and principles of the original EGD agenda.

The evolution of the EU's climate agenda in its broader political context is also mirrored in the Commission AWPs covered in the previous survey and summarized in Table 3.2 below.

The sequence of AWPs creates continuity by presenting the same structure of an opening section followed by discussion of the six headline goals, the first being the EGD agenda, in each of the four documents. Concerning the number and substance of new policy initiatives listed at the end of each AWP, the share of those listed under the heading of the EGD corresponds roughly to its consideration as one of six headline goals with no evident change over time.[2]

To summarize, the data reviewed in this section contrast two dynamics of agenda-setting at two different institutional levels: first,

Table 3.2 The European Green Deal agenda in the EU Commission Annual Work Programs

	AWP 2020	*AWP 2021*	*AWP 2022*	*AWP 2023*
Overall EU ambitions and targets	Transformative challenges post-crisis: fair, digital, climate-neutral Europe in a volatile world	Foundation for systemic change: new industrial strategy and green transition recovery	Recovery from crisis and transition to socially fair and climate-neutral EU through RRF	War confirms case for collective effort to achieve transformation; defense of values and democracy
Key priorities of EU climate action	EGD as growth strategy; adoption of Climate Law and circular economy action plan	Launch of Fit for 55 and CBAM proposal; circular economy plan and farm to fork strategy	Launch of Green Bonds and carbon removal certificates; socially fair energy and green transition	Swift adoption of Fit for 55 package; electricity market reform and hydrogen strategy

Sources: EU Commission AWPs 2020–2023, listed in the Appendix.

at the highest macropolitical level of the EU, clear shifts of attention that emerge particularly in conclusions of the European Council; and second, at the meso-political level, an accommodation of exogenous shocks through a more continuous adjustment of agendas by the Commission that combines an expansion in the scope and policy instruments of EU climate action with the reaffirmation of its defining goals and principles. In a nutshell, this goal consists of the commitment to achieving carbon neutrality by mid-century and the definition of the EGD agenda as the new model for growth and competitiveness of the EU.

3.2 Policy venues: expansion and differentiation of the European Green Deal governance framework

As a consequence of its expanding agenda, the institutional framework for EU climate action has encountered pressure for change in the period since the adoption of the EGD. The theoretical perspective of PET suggests two mechanisms through which a change of policy images affects policy venues: in a vertical dimension, through an intervention by macropolitical institutions into extant policy monopolies through the assignment of decision-making roles and competences to new venues and subsystems; and on a horizontal level, through effects of competition and friction between subsystems charged with related but distinct tasks of policy-making. Both mechanisms deserve consideration in the present case, as the overall EGD governance framework has expanded under the direction of EU executive institutions and must accommodate the interaction of multiple related decision-making processes.

The main pressure for innovation for the climate policy subsystem is a dual task of expansion: namely, the creation of policy instruments beyond the realm of regulatory policy-making towards the distribution of funds and stronger mechanisms of coordinative governance; and the adjustment of governance frameworks to cover the EU's proclaimed geopolitical targets and connect instruments of internal climate governance with its external dimension. As the subsequent reconstruction will show, this dynamic of expansion has evolved through a logic of incremental functional differentiation: instead of disrupting the policy subsystem of venues and decision-making processes identified with EU climate action, additional subsystems have been created with separate and distinct competences and decision-making. These are connected to one another through coordination and leadership by the Commission but remain shielded from one another with regard to potential political intervention or veto options (cp. Wendler 2023, von Homeyer et al. 2022).

Within the EGD governance framework, three distinct policy subsystems can be distinguished that are distinct from one another with regard to their composition, degree of supranational competence and decision-making procedures:

(1) "Core" climate policy subsystem: from its establishment, EU climate policy has primarily been pursued through the adoption of regulatory legislation to restrict and price carbon emissions and create rules for standard-setting, particularly in the fields of energy production and consumption, product and energy efficiency standards, buildings, transport and agriculture. Key components of the regulatory framework are market-based instruments such as emissions trading; rules applying to Member States such as effort-sharing regulations; and more sector-specific rules for renewable energy and efficiency standards for buildings, vehicles, fuels and industry products. Policy developments in this core subsystem have evolved since the 2000s, the starting point being defined primarily by the establishment of the first pilot phase of emissions trading in 2005. In institutional terms, a clearly defined and stable subset of venues and decision-making procedures can be identified that create the foundation for this subsystem: these include the institutional triangle between the Commission and its Directorate-General (DG) for Climate, the Council of the EU primarily in its configuration for the environment (ENV) and energy (TTE), and the EP and particularly its committees for the environment (ENVI) and industry and competitiveness (ITRE). Based on formal rules of the Ordinary Legislative Procedure (OLP), the key mode of interaction between these venues is the trilogue procedure as a relatively informal and cooperative route of decision-making (Brandsma et al. 2021, Bressanelli et al. 2016, Roederer-Rynning 2015, Rosen 2016), used to negotiate a consensus on virtually any piece of legislation in the realm of energy and climate policy. The assignment of legislative proposals to committees and individual rapporteurs in the EP varies according to the proximity of legislation to aspects of industrial and economic regulation, ecological aspects and questions of land use and agriculture.

Nevertheless, this core subsystem is close to what would be considered a policy monopoly from the perspective of PET, particularly because of the significant role of the Commission as the initiator of legislative proposals, the strong role of EP committees for negotiating proposals and frequently the long-term experience of involved rapporteurs, and the high degree of formalization and result-oriented dynamic of inter-institutional bargaining through trilogue. Considering the long-term presence of this framework, the relative political leadership role of the EP for the development of EU climate policy (Buzogany 2021, Wendler 2019, Burns 2017), it seems appropriate to denominate it as the "core" subsystem for EU climate policy.

(2) "Smart investment" mechanism through executive cooperation: the creation of the RRF as the main vehicle for the disbursement of loans and grants to Member States under the NGEU and REPEU programs creates an institutional innovation that is, however, separate from the established core climate policy subsystem. The main link between this framework and EU climate action is established through the reference to the green transition as the first of six pillars for the scope of application in the EU Regulation establishing the RRF (Article 3 of Regulation 2021/241 of February 12, 2021). An additional key element of guidance is the provision within the same regulation that at least 37% of funds allocated through RRF plans as requested by Member States should contribute to the green transition and issues of biodiversity as specified through a methodology for climate tracking set out in the annex of that regulation (Recital 23). Concerning its governance and oversight mechanism, however, the RRF mechanism is designed as a procedure of executive cooperation between Member States and the Commission; the EP is involved through a regular reporting procedure but without rights of intervention into the approval of RRF plans and disbursement of funds as such. Within the Commission, the steering role is assigned to a task force within the Secretariat-General in cooperation with the DG for Economic and Financial Affairs in the Commission and therefore to units outside the established realm of EU climate action.

This aspect also applies to the process of creating the RRF governance mechanism. Here, the political agreement at the level of the European Council concerning the overall size of funds and share of grants as compared to loans was transposed through a legislative procedure based on a Commission proposal for the regulation specifying the rules for allocation and oversight. The primary involved EP committees were those for the Budget and Economic and Financial Affairs, concluding negotiations with the Commission and Council through trilogue. Under its provisions for general and specific objectives of the RRF mechanism, the regulation establishes the requirement to use the procedure in compliance with both the EU's 2030 interim climate targets and the objective of climate neutrality by 2050. In short, in legal and institutional terms, the RRF mechanism is established in accordance with but separate to the core EU climate policy process described above.

With the launch of the REPEU agenda to achieve independence from Russian fossil fuels, the RRF governance mechanism has been partially adjusted using the same legislative route and configuration of committees as in the original regulation based on the OLP. With the requirement to Member States to include specific REPEU chapters into their recovery and resilience plans, additional criteria are added, but the overall target of 37% green investment is retained (EPRS 2023, Wendler 2023); no substantial change has been introduced into the mechanism through which plans are reviewed and approved. In institutional terms, the main innovation brought at this stage is the funneling of additional funds into the RRF framework, tapping into remaining reserves of RRF funds and voluntary transfers from other EU budgetary tools. The only direct link between the recovery mechanism and established tools of EU climate governance is the sale of CO_2 certificates from the Market Stability Reserve of the EU emissions trading system. Arguably, this transfer presents the risk of giving up an important instrument for controlling the price of certificates and adding downward pressure on those prices to raise an additional sum of €20bn. At the level of policy, the inclusion of exemption clauses from the "Do no significant harm" principle to allow investment in infrastructure for fossil fuel supplies to meet short-term shortages is a second point that potentially compromises goals of the EGD agenda. Concerning the establishment and evolution of policy venues, however, the operation of the RRF combines a high degree of legal formalization and supranational competence of the Commission with a clear separation from venues and decision-making of the regulatory core of EU climate action.

(3) External action and governance: the external dimension of EU climate action lacks a unified institutional framework and is comprised of a range of governance mechanisms mostly without legally binding quality. In this respect it mirrors other fields of EU policy-making where external action is governed through fewer supranationalized procedures and instruments than internal action. Concerning extant regulatory action, the CBAM proposed in the initial EGD agenda is the most relevant component in combination with EU internal legislation that has either explicit or de facto external effects, such as emissions trading for shipping and aviation. In particular, the more recent policy packages in the REPEU agenda include elements of external action that are enacted in part through the EU's external relations policy framework under the coordination of the European External Action Service or through specifically designed frameworks; concerning the latter, the energy platform to monitor and coordinate LNG purchases by Member States and the promotion of an external hydrogen strategy are particularly relevant in this context. With the exception of CBAM, these evolving policy initiatives and governance frameworks are, however, separate from the two policy subsystems reviewed above and not part of its core agendas of legislative revision under "Fit for 55" and the RRF mechanism.

In a combined perspective on these three subsystems, a key question for evaluating dynamics of stability and change is how they have evolved and interacted through the different stages of the EGD agenda discussed here: more specifically, how different components of the policy programs discussed here are assigned to the subsystems and what forms of interaction are created as a consequence, summarized in Table 3.3 below.

Based on this overview, four main factors can be identified that have shaped how the evolution of agendas related to the EGD have been transposed into decision-making and contributed to institutional change of its governance framework.

First, several aspects in the evolution of the EGD indicate a defining role of the core subsystem and legislative policy process for its functioning as a governance framework. The most evident is the adoption of the European Climate Law and its legally binding

Table 3.3 EU climate policy programs and policy subsystems

	Core climate subsystem: regulatory legislation	*Distributive / investment mechanisms*	*External action framework*
EGD (2019)	European Climate Law Fit for 55 package: 19 proposals (1) GHG pricing / restriction (2) Energy efficiency standards (3) Energy production (4) Regulation of fuels (5) Land use / forestry (6) Compensation	Just Transition Fund Social Climate Fund (implemented outside RRF)	(CBAM suggested as option but adopted in core subsystem)
NGEU (2020)	Legal foundations of RRF mechanism	Establishment of RRF mechanism (NGEU)	
REPEU (2022)	Rules for REPEU chapters Adjustment of Fit for 55 (1) Renewable energy (2) Efficiency (buildings, transport)	Reinforcement and adjustment of RRF mechanism (REPEU chapters)	Energy platform
GDIP (2023)	Adjustment of regulation on permitting EU framework for flexibilization of state aid rules	Funding mechanisms proposed but outside of RRF (not yet fully specified)	Trade agenda (not yet fully specified)

Sources: COM 2019, 2022a, 2023a, 2023c, EPRS 2022, 2023a, 2023c, 2023f, 2023g.

commitment of the EU and its Member States to achieving carbon neutrality by mid-century in combination with a prescribed path of oversight and review. As a consequence, a package of legislative proposals under the heading of "Fit for 55" has been processed within the core subsystem to specify the transition to the interim target of emission reductions by 2030 (Perez de las Heras 2022, EPRS 2023b). This process has prompted a decision-making processes about sector-specific adjustment of rules and regulations that has proceeded through the entire span of four agenda-setting stages identified above without substantial modification of its core goals. Through the packaging of proposals in this package and control of the Commission over the legislative process, this centerpiece of the EGD agenda has remained largely shielded from exogenous shock and progressed steadily towards adoption of almost all its components (cp. Wendler 2023). Finally, key political agents of the core policy subsystem assume control over the other subsystems identified through the passage of legislation that defines the operation and specific criteria of "green recovery" programs governed through the RRF mechanism. Passing through the OLP, the design of the framework for dispersing funds to Member States based on green conditionality has been co-decided by the EP and assigns a strong oversight function to the Commission, thereby adding an important supranational component to the establishment as distinguished from the operation of the RRF mechanism.

A second factor is the availability of parallel processing of distinct policy-making demands. The assignment of functionally differentiated types of policy-making – those associated with regulatory legislation, distributive mechanisms and external governance – to separated institutional frameworks and based on different degrees of supranational competence is a typical feature of EU governance. In the case covered here, it is equally present through the clear distinction of the RRF mechanism from decision-making on regulatory policy and supports a form of processing of policy challenges identified in the theoretical language of PET as "parallel processing": the distribution of different policy-making tasks to separate venues and decision-making procedures. As policy-making tasks assigned to the three subsystems are separated, horizontal friction is minimized.

A third aspect is the targeted direction of revisions in policy subsystems by the Commission; this aspect becomes relevant particularly in the adoption of the REPEU agenda, where the whole range of policies comprised in the EGD agenda is reconsidered in a new geopolitical context. Under the direction of the Commission, however, intervention in ongoing policy processes is limited to very few and targeted revisions: in the field of regulatory policy-making, revisions were introduced in a single proposal covering only a more stringent target for the renewable energy directive, while modifications in related fields such as energy efficiency and energy performance of buildings were informally left to legislators to consider in ongoing negotiations (EPRS 2023). Interventions through formal proposals were therefore reduced to a single file. The smart investment pillar of REPEU, in turn, was realized by building on the extant RRF mechanism and limiting modifications primarily to the requirement of adding a dedicated chapter in national recovery plans and mobilizing additional funds. This point adds to previous accounts emphasizing the political leadership of the Commission for progress of EU climate policy (Skjaerseth 2017, Dreger 2014).

Finally, an aspect that contributes to the relative stability of the EGD governance framework is the asymmetrical assignment of supranational competence. The expansion of the EGD policy image is not fully matched by an extension of enduring supranational competence into the corresponding fields of green recovery and external governance. Especially, the mechanisms of joint EU action in its external energy relations and investment largely rely on non-binding mechanisms of cooperation and establish only limited safeguards against unilateral action by Member States. The proclaimed turn of EU climate action towards a more externally oriented, geopolitical approach therefore encounters an institutional barrier through the asymmetrical allocation of supranational competence in favor of internal regulatory action and the core climate subsystem. The main link between both, as discussed above, is the adoption of CBAMs, established as one of the key tools of external climate action but adopted through its established policy-making procedure.

In conclusion, the institutional development of the EU climate governance framework presents itself as a case of resilience and continuity even in the presence of relatively far-reaching change of policy images and overall agenda at the macropolitical level of the EU. At an analytical level, this finding is not entirely surprising when considering the structure of the political system of the EU: namely, a highly differentiated and relatively decentralized framework of policy-making with a strongly cooperative mode of decision-making and inclination towards incremental policy-making and effects of path dependency. On a political level, the finding is more ambivalent: on the one hand, it shows the capacity of EU institutions and particularly the Commission to promote its agenda of decarbonization even in the presence of strong external challenges and its ability to establish new linkages between climate action and policy-making in response to emerging external shocks. At the same time, the proclaimed shift of agendas towards a more strategic and geopolitical orientation so far lacks credibility, as policy-making dynamics and institutional competence remain located primarily in the field of EU internal climate action and its core subsystem of regulatory policy-making.

3.3 Policy feedback: incremental progress towards the net-zero emission target

In this third step, we turn to the evaluation of policy-making results: to what degree the exogenous shocks and agenda changes covered in the previous sections have caused a disruption or policy-making shift for core targets of the EGD agenda. The previous analysis has suggested that even faced with these challenges, EU institutions have responded by evaluating these shocks as opportunities for investment in a sustainable recovery; synergies were emphasized between decarbonization and energy security, thereby reconfirming key targets of the EGD as solutions to perceived challenges. The following steps proceed through the three policy subsystems identified above to evaluate decision-making in this regard.

(1) Regulatory policy-making: based on the foundation of the European Climate Law passed through the legislative route of the OLP and negotiations through trilogue in June 2021, the launch of the Fit for 55 package to enact its interim targets by 2030 has particularly defined legislative processes within this subsystem. Covering the entire scope of the EGD agenda, Fit for 55 is a sprawling collection of 19 legislative proposals covering major challenges of decarbonization such as energy efficiency, renewable energies, regulation for fuels and vehicles, buildings and transport. For reasons of space, Table 3.4 below lists all legislative proposals in six major fields of policy-making, key innovations included in these proposals, and their respective state of adoption at the time of writing. As this survey shows, almost all components of the package have passed through the legislative process using the established trilogue procedure. In terms of policy content, all have resulted in a moderate to substantial reinforcement of decarbonization targets, most notably raising targets for renewable energy from 40% to 42.5% and establishing more stringent goals for energy efficiency (Buzogany et al. 2023, Buzogany & Cetkovic 2021).

(2) Green investment: the allocation of grants and loans to Member States through the RRF mechanism is in its second stage at the time of writing, as recovery plans including a new REPEU chapter were due to be submitted by August 2023. Compared to the initial stage of the EGD agenda, the enactment of the RRF

Table 3.4 Proposals and decision-making on legislation in the Fit for 55 package

EGD component	Legislative acts and key innovations		Decision-making procedure
(1) GHG emission pricing and restriction	(1.1)	ETS: GHG emissions reduction in covered sectors by 63% by 2030; free allowances to be phased out by 2032 as part of CBAM; establishment of ETS II covering buildings and road transport	(1.1) Pp 07-21, TA 12-22, ap 05-23 (1.1b) Pp 05-22, TA 12-22, ap 04-23
	(1.1b)	MSR: auctioning of €20bn allowances to support RRF, proposed as part of REPEU	(1.2) Pp 07-21, TA 11-22, ap 03-23
	(1.2)	ESR: increased 40% GHG reduction targets with increased national contributions; linear emissions trajectories for MSs; reinforced flexibilities (banking / borrowing)	(1.3) Pp 07-21, TA 12-22, ap 05-23
	(1.3)	CBAM: phase-in from October 2023 and full establishment by January 1, 2026; coverage of cement, electricity, fertilizers, iron/steel, aluminum and hydrogen; phase-out of free allocations starting 2026 with fixed rates through 2032	
(2) Energy efficiency	(2.1)	EE directive: reduction of primary and final energy consumption by 11.7% at EU level and set targets of primary and final consumption (992Mtoe / 763 Mtoe, respectively), annual energy savings by 1.5% on average by MSs through 2030, progressing from 1.3% to 1.9% from 2026; reinforced rules for public sector (1.9% annually)	(2.1): Pp 07-21, TA 03-23, ap 10-23 (2.2): Pp 12-21, negotiation ongoing
	(2.2)	Buildings: zero-emission requirements for new buildings by 2030 (public 2027), increasing renovation speed: improvement of energy class G to F and E by 2027 and 2030, respectively in COM proposal; reinforcement and harmonization of classification standards	(2.3): Pp 07-21, TA Oct 22, ap 03-23 (2.4.): Pp 07-23, response by EP and Council pending

(Continued)

Table 3.4 (Continued)

EGD component	Legislative acts and key innovations	Decision-making procedure
	(2.3) Cars / vans: emission cap from 2021 and reductions by 15% annually from 2025 to 2029, 55% for new cars and 50% for new vans by 2030 and 100% by 2035; additional clauses for e-fuel combustion after 2035; exemption clauses only for small-scale producers; regulatory incentive mechanism for producers through end 2029	
	(2.4) Trucks: proposal offers incentives to use low-emission trucks and loosen rules for low-emission trucks; set rules for cross-border traffic, encourage cross-modal transport	
(3) Energy production	(3.1) RE directive: renewable target raised to 42.5% by 2030 with 2.5% indicative top-up, adoption of sectoral targets (industry, transport, buildings, heating and cooling), tightened rules on sustainability of biofuels; accelerated permitting	(3.1): Pp 07-21 amended in 05-22, TA 03-23, ap 09-23
	(3.2) Energy taxation: proposal suggests to tax fuels according to energy content and environmental performance, to create a categorization of fuels, phase-out of exemptions and disallowing tax-free use of fossil fuels for intra-EU transport	(3.2): Pp 07-21, considered by EP and Council but requiring unanimity
	(3.3) Methane: proposal suggests compulsory measurement and verification, improve leak detection and eliminate routine venting; EP calls for setting 2030 EU-wide binding methane targets and national limits	(3.3): Pp 12-21, EP and Council consulted, TA pending

(4) Regulation of fuels	(4.1)	Gas directive: proposal seeks to incorporate and set rules for certification of renewable and low-carbon gases, consumer rights and cross-border EU hydrogen networks; EP stresses need to support hydrogen corridors requested in REPEU	(4.1): Dir.: Pp 12-21, EP and Council consulted and TA pending
	(4.1b)	Gas regulation: proposal seeks to incorporate and cover rulework for hydrogen and renewable gases, set up legal framework for hydrogen network; EP position adopts REPEU targets on sustainable biomethane to replace 20% of Russian gas supplies; coordinate gas purchases by MSs; Council encourages measures to limit access to Russian gas	(4.1b): Pp 21-21, EP and Council consulted and TA pending
	(4.2)	Maritime fuels: reduction of GHG intensity of fuels used by shipping decreasing, starting at 2% annually to 80% by 2050; application regardless of flag with exemptions for specific vessel types (warships, fishing, etc.); sub-target of 2% renewable non-biofuels from 2034; use of on-shore energy	(4.2): Pp 07-21, TA 03-23, ap 09-23
	(4.3.)	Aviation fuels: rules for uptake of sustainable aviation fuels, starting from 2025 at 2% and increasing every five years (6–20–34–42–70%) by 2050; additional proportion must comprise synthetic fuels (35% by 2050); exclusion of fuels from food or feed crops, palm and soya; voluntary labeling	(4.3.): Pp 07-21, TA 04-23, ap 10-23
	(4.4.)	Alternative fuels infrastructure: deployment targets for transport; recharging pools every 60km along TEN-T network by 2025; charging station for electric heavy goods, every 60 km by 2030; hydrogen refueling stations from 2030 every 200 km along TEN-T network; rules for airports providing electricity to stationary aircraft from 2025	(4.4.): Pp 07-21, TA 03-23, ap 09-23

(Continued)

Table 3.4 (Continued)

EGD component	Legislative acts and key innovations	Decision-making procedure
(5 Land use and agriculture	(5.1.) LULUCF directive: objective of 310mn t CO_2 removals by 2030 with binding national targets; flexibility rules for trading surpluses and provisions for natural disturbances; mechanisms for corrective action; improved verification (5.2.) EU Forest Strategy: diverse range of points, including related legislation (construction products, ecosystem restoration, forest reproductive material directives) and EP resolution calling for sustainable forestry and role of wood for replacing energy-intensive materials	(5.1.): Pp 07-21, TA 11-22, ap 04-23 (5.2.): Com strategy 07-21 and EP OI procedure 02-22
(6) Compensation mechanisms	(6.1.) Social Climate Fund: establishment of SCF from 2026 to 2032 to help vulnerable households, micro-enterprises and transport users to cope with price impacts of carbon pricing; SCF is created as part of the EU budget filled by external assigned revenues, mostly ETS allowances and Member State co-financing of 25%, up to €65bn; introduction of ceiling of 37.5% to finance temporary direct income support under national climate plans submitted by MSs to COM	(6.1.): Pp 07-21, TA 21-22, ap 05-23

List of abbreviations: (1) Legislative acts: ETS = Emissions Trading System; MSR = Market Stabilization Reserve; ESR = Effort Sharing Regulation; EE = Energy Efficiency; RE = Renewable Energies; LULUCF = Land Use and Land Use Change Framework; (2) Decision-making procedure: Date and month of legislative proposal (Pp); of agreement reached in trilogue (TA); and of formal adoption of legislation (ap).
Sources: EPRS 2023b and EP Legislative Train Schedule (www.europarl.europa.eu/legislative-train/theme-a-european-green-deal).

and allocation of funds signifies major policy change as it greatly increases the size of financial resources mobilized within the remit of EU climate action, in addition to creating a new set of instruments for the release of bonds and borrowing of debt by the EU. Particularly from the perspective of PET and its frequent application to the evolution of budget policies and spending, the onset of the Covid pandemic can therefore be reconstructed as a source of significant policy change resulting in new forms of resource mobilization through issuing of Green Bonds and a more strategic set of funding mechanisms. From the more specific perspective on the evolution of EU climate action and the core goals of the EGD agenda taken here, however, it becomes clear how the policy change brought about by the adoption of the RRF is additive to and separate from the ongoing evolution of the regulatory framework of EU climate action.

The specific effects of funds disbursed by the EU through the RRF mechanism are amalgamated with economic and structural policies of Member States; a detailed evaluation is beyond the scope of this present volume due to the wide range of settings and policies affected and time horizons involved in the implementation of Member State action plans. The main aspect to be highlighted here, however, is that the activation of the RRF mechanism has not led to a noticeable shift or slowdown of core regulatory policies launched through the initial EGD agenda and pursued through the legislative package reviewed in detail above. For the prescription of reductions in carbon emissions, the regulatory framework of legislation remains the main driver of change, whereas the allocation of funds through the RRF mechanism is one of facilitation across a range of policy fields and applied through mechanisms of multi-level governance. In this context, a critical test for detecting possible policy shifts or reversals is the adoption of the REPEU package, particularly through its provisions to allow Member States to invest in infrastructure for the diversification of energy supplies, including those based on fossil fuels, and particularly a widening of LNG supplies. The relevant provisions of the regulatory framework, however, establish relatively strict limits for any deviation from the "Do no significant harm" principle (Wendler 2023, EPRS 2023a,f, Buzogany et al. 2023). Based on this supranational guidance, extant studies have identified partial modifications but no policy reversal through investment clauses in the REPEU regulations.

(3) External action: within this dimension of EU climate action, a more diverse and potentially incoherent set of policy-making processes is observed that forms part of efforts to realize the EU's shift towards a more geopolitical approach to climate action. Three main components can be distinguished. The first consists of efforts to define a strategic approach to external energy relations of the EU, particularly through the coordination of LNG purchases through the energy platform and the adoption of an external energy security strategy. The second aspect is the inclusion of climate conditionality criteria in all EU external relations, most importantly in the conclusion of trade agreements. In this regard, current negotiations with Mercosur, but also with Indonesia and Australia, particularly stand out as important for considering aspects of land use, agriculture and critical materials required for the energy transition. While this aspect receives growing attention in extant research on the negotiation of free trade agreements, at this point it seems implausible to recognize green conditionality in trade agreements as a source of major breakthroughs of EU climate action on a global scale. Finally, a recent and increasingly defining aspect of the EU's external climate action is the strategic engagement with the US following the adoption of industrial policy legislation, covered in more detail in the next chapter. Besides direct engagement with the US counterpart through the creation of a task force, the launch of the GDIP agenda establishes the response by the EU to the more active and protectionist policies promoted through the bills endorsed by the Biden administration. Actual policy progress on this agenda, however, remains partial and elusive at the time of writing, not least due to considerable disagreement between Member States on the contours of a future EU industrial policy.

In addition to these new mechanisms, the establishment of a CBAM by the EU is a major step towards a more effective external dimension of the EGD (Evans et al. 2021). Its creation, however, confirms the relevance of the core regulatory subsystem rather than new governance mechanisms: Adopted as an EU Regulation in May 2023, CBAM is part of the original EGD agenda and was negotiated as part of Fit for 55, rather than resulting from more recent initiatives to strengthen the geopolitical presence of the EU.

Summing up, the policy-making process comprised in the EGD agenda has expanded to include additional resources and instruments but evolves in a path of continuity towards the target of carbon neutrality even in the presence of several subsequent events of exogenous shock. Comparing between policy subsystems, the clearest case of policy stability is the stepwise revision and extension of the regulatory framework that establishes the foundation of EU climate action. The major policy innovation in the period covered is the mobilization of resources for green investment through the RRF mechanism which, however, has also been only partially adjusted and reinforced rather than reversed through the external shock of war in Ukraine. No major aspect of policy processes covered here points to a reversal of zero-carbon targets caused by exogenous shocks. On the contrary, the Russian attack on Ukraine, in particular, has prompted a policy-making response by the EU that has resulted in a moderate increase in the stringency of policy targets and regulatory legislation, even when considering exemption clauses within the REPEU program that compromise EGD targets such as investment in diversification of fossil fuel supplies.

3.4 Discussion: vectors of stability and punctuation in the European Green Deal governance process

To summarize the findings of this first case study, we return to the theoretical framework presented at the outset and the indicators defined there to identify vectors of stability and change in the evolution of climate governance frameworks. Vectors, as discussed above, are defined as a term for the observation of change or continuity of these indicators in the sequence of the different stages of agenda-setting and policy-making within the EGD governance framework as reviewed here. From this point of departure, the main findings of the case study can be summarized as follows:

• Concerning its *policy image*, the main vector of change in the EGD governance process concerns the *scope* of challenges addressed and fields of action covered by EU climate action: from the primarily regulatory approach of the original EGD to a broader range of policy-making instruments including modes of distributive and external governance and a more strategic geopolitical orientation. A stable core of policy ideas, however, is retained through the commitment to achieving net-zero carbon emissions and the definition of core EGD policies as a model of sustainable growth for the EU; both aims confirm the *priority* of climate targets and definition of *transformative dynamics* based on a clean energy transition and decarbonization of industry and society. A key factor in this regard is the role of the Commission as an agenda-setter at the meso level between the more volatile definition of topics by the European Council and specialized policy subsystems.

- As a consequence of broader ambitions expressed in agenda-setting, the *policy venues* involved in EU climate action have evolved into a set of functionally differentiated subsystems primarily to cover its regulatory, distributive and external policy-making components. As shown above, however, the core policy subsystem for regulatory policy shows a great degree of policy stability; in addition, the RRF mechanism as the main vehicle for distributive components of current EU climate action has been created through an institutional design with a strong role of oversight of the Commission and formalized through the established route of legislative decision-making through the OLP. As a consequence, *vertical direction* is exerted by the Commission in a targeted way that has limited adjustments and interventions into ongoing policy-making and thereby ensured stability, particularly in the adoption of the Fit for 55 package. *Horizontal friction* between subsystems has been avoided through institutional separation, while informal negotiation mechanisms through trilogue minimize this factor within regulatory policy-making. Public visibility and *accountability* has spiked at several occasions relating to the operation of RRF mechanisms but remains low with regard to regulatory policy in the core policy subsystem, mainly due to the informal character of legislative agreement through trilogue.

- As a result, *policy-making feedback* has evolved with continuity through the different stages of the EGD agenda, especially at the level of regulatory policy. The envisaged *speed* of policies towards net-zero decarbonization has been maintained and even moderately increased through the adoption of more stringent legislation, particularly in the areas of renewable energy and efficiency targets; in addition, the targeting of recovery funds on projects related to green investment provides additional support for an acceleration of decarbonization at Member State level. The *stringency* of EU climate action is upheld, as the foundation of the EGD in a regulatory framework of carbon pricing and reduction has not been abandoned and has even been extended to the external dimension through the adoption of a CBAM. The achieved degree of *irreversibility* of EU climate policies is high, especially in the regulatory field, as a repeal or fundamental revision of EU legislation would require strong legislative majorities and would need to accommodate political and legal challenges based on commitments in the European Climate Law. Possible reversals are more likely at the level of distributive policies when the RRF mechanism is set to expire in 2026 and if no successor mechanism is created.

3.5 Conclusion: policy-making stability at the expense of dynamism?

To summarize, the present case study has shown that policy-making under the EGD agenda has evolved under pressures for change from a volatile external environment, as reflected in agenda-setting dynamics at the macropolitical level of the EU. In response to the exogenous shocks of the Covid pandemic and war in Ukraine, macropolitical institutions of the EU have expanded the original focus on achieving decarbonization through regulatory policy to include more active and targeted investment in green infrastructure and adopt a more geopolitical approach to questions of energy security, efficiency and renewable sources.

The transposition of this ambitious agenda into policy-making, however, highlights elements of PET that explain an evolution of relative policy-making stability. Rather than prompting a shift or disruption of extant policy venues towards the inclusion of new agents and interest coalitions, the EGD governance process has unfolded through the addition of new governance frameworks that leave the core subsystem of EU regulatory climate policy largely unaffected. Within this core, policy stability is maintained through the incremental revision of legislation to include more stringent rules for emission reductions even in the face of adverse exogenous shocks. More recent efforts to respond to the rapidly changing geopolitical environment of the EU and strengthen its external action capacity so far largely fall short of prompting a major policy-making breakthrough; this aspect demonstrates a degree of inertia caused by the continuity and incrementalism of EU climate policy-making. Evaluated in political rather than theoretical terms, the finding of policy stability therefore has ambivalent implications for the EU as an actor of climate governance: namely, as a supporting factor for continued progress towards the goal of decarbonization but also an obstacle against more decisive change in the strategic direction of the overall EU climate governance framework.

Notes

1 The 2019 Speech by President-elect von der Leyen in the EP Plenary on the occasion of the presentation of her College of Commissioners and their programme was included in this survey as it represents a similarly broad thematic outlook on the political agenda and priorities of the incoming Commission.
2 In each AWP where policy initiatives are listed, the quantitative share of policy initiatives under the EGD is roughly one-sixth of all listed initiatives: the AWP for 2020 lists 8 out of 43 new policy initiatives under the heading of the EGD and 4 out of 44 for the year 2021 (with the Fit for 55 package and its 13 sub-points listed as a single initiative). For the year 2022, this share is 5 out of 32 and for the year 2023, the AWP lists 9 out of 43 for the EGD.

References

Anghel, Veronica, and Erik Jones. 2023. "Is Europe Really Forged Through Crisis? Pandemic EU and the Russia–Ukraine War." *Journal of European Public Policy* 30(4): 766–86. doi:10.1080/13501763.2022.2140820.
Boasson, Elin Lerum, and Jørgen Wettestad. 2013. *EU Climate Policy: Industry, Policy Interaction and External Environment*. Farnham [u.a.]: Ashgate.
Bonciu, Florian. 2022. "The Implications of the REPowerEU Plan in Accelerating the Implementation of the European Union's Hydrogen Strategy." *Romanian Journal of European Affairs* 22(2): 100–14.
Bongardt, Annette, and Francisco Torres. 2022. "The European Green Deal: More than an Exit Strategy to the Pandemic Crisis, a Building Block of a Sustainable European Economic Model*." *JCMS: Journal of Common Market Studies* 60(1): 170–85. doi:10.1111/jcms.13264.
Brandsma, Gijs Jan, Justin Greenwood, Ariadna Ripoll Servent, and Christilla Roederer-Rynning. 2021. "Inside the Black Box of Trilogues: Introduction to the Special Issue." *Journal of European Public Policy* 28(1): 1–9. doi:10.1080/13501763.2020.1859600.
Bressanelli, Edoardo, Christel Koop, and Christine Reh. 2016. "The Impact of Informalisation: Early Agreements and Voting Cohesion in the European

Parliament." *European Union Politics* 17(1): 91–113. doi:10.1177/ 1465116515608704.

Bressanelli, Edoardo, and David Natali, eds. 2023. "Governing the EU Polycrisis: Institutional Change After the Pandemic and the War in Ukraine." *Politics and Governance* Open Access 11(4). doi:10.17645/pag. i374.

Burns, Charlotte. 2017. "The European Parliament and Climate Change: A Constrained Leader?" In *The European Union in International Climate Change Politics*, eds. Rüdiger Wurzel and James Connelly, Routledge Studies in European Foreign Policy, London: Routledge, 52–65. doi:10.4324/9781315627199-16.

Burns, Charlotte. 2019. "In the Eye of the Storm? The European Parliament, the Environment and the EU's Crises." *Journal of European Integration* 41(3): 311–27. doi:10.1080/07036337.2019.1599375.

Buti, Marco, and Sergio Fabbrini. 2023. "Next Generation EU and the Future of Economic Governance: Towards a Paradigm Change or Just a Big One-Off?" *Journal of European Public Policy* 30(4): 676–95. doi:10.1080/ 13501763.2022.2141303.

Buzogány, Aron, and Stefan Ćetković. 2021. "Fractionalized but Ambitious? Voting on Energy and Climate Policy in the European Parliament." *Journal of European Public Policy* 28(7): 1038–56. doi:10.1080/ 13501763.2021.1918220.

Buzogány, Aron, Stefan Ćetković, and Tomas Maltby. 2023. "EU Renewable Energy Governance and the Ukraine War: Moving Ahead Through Strategic Flexibility?" *Politics and Governance* 11(4): 263–74. doi:10.17645/pag. v11i4.7361.

COM (European Commission). 2019. "The European Green Deal." 640 final, Brussels, 11.12.2019. https://eur-lex.europa.eu/resource.html?uri= cellar:b828d165-1c22-11ea-8c1f-01aa75ed71a1.0002.02/DOC_1&for mat=PDF.

COM (European Commission). 2020. "Europe's Moment: Repair and Prepare for the Next Generation." https://eur-lex.europa.eu/legal-content/EN/TXT/ PDF/?uri=CELEX:52020DC0456.

COM (European Commission). 2021. "Recovery and Resilience Scoreboard. Thematic Analysis: Clean Power." https://ec.europa.eu/economy_finance/ recovery-and-resilience-scoreboard/assets/thematic_analysis/1_Clean.pdf.

COM (European Commission). 2022a. "REPowerEU: Joint European Action for More Affordable, Secure and Sustainable Energy." https://eur-lex.eur opa.eu/resource.html?uri=cellar:71767319-9f0a-11ec-83e1-01aa75ed7 1a1.0001.02/DOC_1&format=PDF.

COM (European Commission). 2022b. "REPowerEU Plan." https://eur-lex. europa.eu/resource.html?uri=cellar:fc930f14-d7ae-11ec-a95f-01aa75ed7 1a1.0001.02/DOC_1&format=PDF.

COM (European Commission). 2023a. "A Green Deal Industrial Plan for the Net-Zero Age." https://eur-lex.europa.eu/legal-content/EN/TXT/PDF/?uri=CELEX:52023DC0062.

COM (European Commission). 2023b. "Launch of the US–EU Task Force on the Inflation Reduction Act." https://ec.europa.eu/commission/presscorner/detail/en/statement_22_6402.

COM (European Commission). 2023c. "Proposal for a Regulation of the European Parliament and of the Council on Establishing a Framework of Measures for Strengthening Europe's Net-Zero Technology Products Manufacturing Ecosystem (Net Zero Industry Act)." https://eur-lex.eur opa.eu/resource.html?uri=cellar:6448c360-c4dd-11ed-a05c-01aa75ed7 1a1.0001.02/DOC_1&format=PDF.

Delbeke, Jos, and Peter Vis. 2015. *EU Climate Policy Explained.* Abingdon, OX [u.a.]: Routledge.

Delbeke, Jos, and Peter Vis. 2019. *Towards a Climate-Neutral Europe: Curbing the Trend.* London: Routledge.

Domorenok, Ekaterina, and Paolo Graziano. 2023. "Understanding the European Green Deal: A Narrative Policy Framework Approach." *European Policy Analysis* 9(1): 9–29. doi:10.1002/epa2.1168

Dreger, Jonas. 2014. *The European Commission's Energy and Climate Policy: A Climate for Expertise?* Houndsmills [u.a.]: Palgrave Macmillan.

Dupont, Claire. 2016. *Climate Policy Integration into EU Energy Policy.* Abingdon, OX [u.a.]: Routledge.

Dupont, Claire, and Diarmuid Torney. 2021. "European Union Climate Governance and the European Green Deal in Turbulent Times." *Politics and Governance* 9(3): 312–15. doi:10.17645/pag.v9i3.4896.

Dupont, Claire, and Sebastian Oberthür. 2012. "Insufficient Climate Policy Integration in EU Energy Policy: The Importance of the Long-Term Perspective." *Journal of Contemporary European Research* 8(2). doi:10.30950/jcer.v8i2.474.

Eckert, Eva, and Oleksandra Kovalevska. 2021. "Sustainability in the European Union: Analyzing the Discourse of the European Green Deal." *Journal of Risk and Financial Management* 14(2): 80. doi:10.3390/jrfm14020080.

Eckert, Sandra. 2021. "The European Green Deal and the EU's Regulatory Power in Times of Crisis." *JCMS: Journal of Common Market Studies* 59(S1): 81–91. doi:10.1111/jcms.13241.

Ekerbout, Milan, Christian Egenhofer, Jorge Nunez Ferrer, Mihnea Catuti, Irina Kustova, and Vasileos Rizos. 2020. "The European Green Deal After Corona: Implications for EU Climate Policy."

EPRS (European Parliament Research Service). 2022. "Briefing: Next Generation EU Delivery: Energy Policy in the National Recovery and Resilience Plans." www.europarl.europa.eu/RegData/etudes/BRIE/2022/738194/EPRS_BRI(2022)738194_EN.pdf.

EPRS (European Parliament Research Service). 2023a. "At a Glance – Agreement on REPowerEU Chapters in Recovery and Resilience Plans." www.europarl.europa.eu/RegData/etudes/ATAG/2023/739330/EPRS_ATA(2023)739330_EN.pdf.

EPRS (European Parliament Research Service). 2023b. "Briefing: Review of the EU ETS 'Fit for 55' Package." www.europarl.europa.eu/RegData/etudes/BRIE/2022/698890/EPRS_BRI(2022)698890_EN.pdf.

EPRS (European Parliament Research Service). 2023c. "EU's Response to the US Inflation Reduction Act (IRA)." www.europarl.europa.eu/RegData/etudes/IDAN/2023/740087/IPOL_IDA(2023)740087_EN.pdf.

EPRS (European Parliament Research Service). 2023d. "EU–US Climate and Energy Relations in Light of the Inflation Reduction Act." www.europarl.europa.eu/RegData/etudes/BRIE/2023/739300/EPRS_BRI(2023)739300_EN.pdf.

EPRS (European Parliament Research Service). 2023e. "EU–US Trade and Technology Council Modest Progress in a Challenging Context." www.europarl.europa.eu/RegData/etudes/BRIE/2023/739336/EPRS_BRI(2023)739336_EN.pdf.

EPRS (European Parliament Research Service). 2023f. "Governance and Oversight of the Recovery and Resilience Facility." www.europarl.europa.eu/RegData/etudes/BRIE/2023/747883/EPRS_BRI(2023)747883_EN.pdf.

EPRS (European Parliament Research Service). 2023g. "Net-Zero Industry Act." www.europarl.europa.eu/RegData/etudes/BRIE/2023/747903/EPRS_BRI(2023)747903_EN.pdf.

Evans, Stuart, Michael A. Mehling, Robert A. Ritz, and Paul Sammon. 2021. "Border Carbon Adjustments and Industrial Competitiveness in a European Green Deal." *Climate Policy* 21(3): 307–17. doi:10.1080/14693062.2020.1856637.

Hodson, Dermot, and Uwe Puetter. 2019. "The European Union in Disequilibrium: New Intergovernmentalism, Postfunctionalism and Integration Theory in the Post-Maastricht Period." *Journal of European Public Policy* 26(8): 1153–71. doi:10.1080/13501763.2019.1569712.

von Homeyer, Ingmar, Sebastian Oberthür, and Andrew J. Jordan. 2021. "EU Climate and Energy Governance in Times of Crisis: Towards a New Agenda." *Journal of European Public Policy* 28(7): 959–79. doi:10.1080/13501763.2021.1918221.

von Homeyer, Ingmar, Sebastian Oberthür, and Claire Dupont. 2022. "Implementing the European Green Deal During the Evolving Energy Crisis." *JCMS: Journal of Common Market Studies* 60(S1): 125–36. doi:10.1111/jcms.13397.

Kreienkamp, Julia, Tom Pegram, and David Coen. 2022. "Explaining Transformative Change in EU Climate Policy: Multilevel Problems, Policies, and Politics." *Journal of European Integration* 44(5): 731–48. doi:10.1080/07036337.2022.2072838.

Lenschow, Andrea, and Viviane Gravey, eds. 2021. *Environmental Policy in the EU: Actors, Institutions and Processes.* Abingdon, OX [u.a.]: Routledge. www.routledge.com/Environmental-Policy-in-the-EU-Actors-Instituti ons-and-Processes/Jordan-Gravey/p/book/9781138392168 (February 22, 2024).

Machin, Amanda. 2019. "Changing the Story? The Discourse of Ecological Modernisation in the European Union." *Environmental Politics* 28(2): 208–27. doi:10.1080/09644016.2019.1549780.

Oberthür, Sebastian, and Ingmar von Homeyer. 2023. "From Emissions Trading to the European Green Deal: The Evolution of the Climate Policy Mix and Climate Policy Integration in the EU." *Journal of European Public Policy* 30(3): 445–68. doi:10.1080/13501763.2022.2120528.

Ossewaarde, Marinus, and Roshnee Ossewaarde-Lowtoo. 2020. "The EU's Green Deal: A Third Alternative to Green Growth and Degrowth?" *Sustainability* 12(23): 9825. doi:10.3390/su12239825.

Pérez de las Heras, Beatriz. 2022. "The 'Fit for 55' Package: Towards a More Integrated Climate Framework in the EU." *Romanian Journal of European Affairs* 22(2). www.ceeol.com/search/journal-detail?id=2549.

Roederer-Rynning, Christilla, and Justin Greenwood. 2015. "The Culture of Trilogues." *Journal of European Public Policy* 22(8): 1148–65. doi:10.1080/13501763.2014.992934.

Rosamond, Jeffrey, and Claire Dupont. 2021. "The European Council, the Council, and the European Green Deal." *Politics and Governance* 9(3): 348–59. doi:10.17645/pag.v9i3.4326.

Rosén, Guri. 2016. "A Match Made in Heaven? Explaining Patterns of Cooperation Between the Commission and the European Parliament." *Journal of European Integration* 38(4): 409–24. doi:10.1080/ 07036337.2016.1141903.

Ryner, J. Magnus. 2023. "Silent Revolution/Passive Revolution: Europe's COVID-19 Recovery Plan and Green Deal." *Globalizations* 20(4): 628–43. doi:10.1080/14747731.2022.2147764.

Samper, Juan Antonio, Amanda Schockling, and Mine Islar. 2021. "Climate Politics in Green Deals: Exposing the Political Frontiers of the European Green Deal." *Politics and Governance* 9(2): 8–16. doi:10.17645/pag. v9i2.3853.

Schramm, Lucas, Ulrich Krotz, and Bruno De Witte. 2022. "Building 'Next Generation' After the Pandemic: The Implementation and Implications of the EU Covid Recovery Plan." *JCMS: Journal of Common Market Studies* 60(S1): 114–24. doi:10.1111/jcms.13375.

Schunz, Simon. 2022. "The 'European Green Deal' – a Paradigm Shift? Transformations in the European Union's Sustainability Meta-Discourse." *Political Research Exchange* 4(1): 2085121. doi:10.1080/ 2474736X.2022.2085121.

Siddi, Marco. 2020. "The European Green Deal: Assessing Its Current State and Future Implementation." www.fiia.fi/wp-content/uploads/2020/05/wp1 14_european-green-deal.pdf.

Siddi, Marco. 2021. "Coping with Turbulence: EU Negotiations on the 2030 and 2050 Climate Targets." *Politics and Governance* 9(3): 327–36. doi:10.17645/pag.v9i3.4267.

Siddi, Marco, and Federica Prandin. 2023. "Governing the EU's Energy Crisis: The European Commission's Geopolitical Turn and Its Pitfalls." *Politics and Governance*, this Special Issue.

Skjaerseth, Jon Birger. 2017. "The Commission's Shifting Climate Leadership: From Emissions Trading to Energy Union." In *The European Union in International Climate Change Politics: Still Taking a Lead?*, eds. Rüdiger Wurzel, James Connelly, and Duncan Liefferink. Abingdon: Routledge, 37–51.

Skjaerseth, Jon Birger. 2021. "Towards a European Green Deal: The Evolution of EU Climate and Energy Policy Mixes." *International Environmental Agreements: Politics, Law and Economics* 21(1): 25–41. doi:10.1007/s10784-021-09529-4.

Skjaerseth, Jon Birger, Peer Ove Eikeland, Lars Gulbrandsen, and Torbjorg Jevnaker. 2017. *Linking EU Climate and Energy Policies: Decision-Making, Implementation and Reform.* Cheltenham: Edward Elgar.

Szulecki, Kacper, Severin Fischer, Anne Therese Gullberg, and Oliver Sartor. 2016. "Shaping the 'Energy Union': Between National Positions and Governance Innovation in EU Energy and Climate Policy." *Climate Policy* 16(5): 548–67. doi:10.1080/14693062.2015.1135100.

Vezzoni, Rubén. 2023. "Green Growth for Whom, How and Why? The REPowerEU Plan and the Inconsistencies of European Union Energy Policy." *Energy Research & Social Science* 101: 103134. doi:10.1016/j.erss.2023.103134.

Wendler, Frank. 2019. "The European Parliament as an Arena and Agent in the Politics of Climate Change: Comparing the External and Internal Dimension." *Politics and Governance* 7(3): 327–38. doi:10.17645/pag.v7i3.2156.

Wendler, Frank. 2022. *Framing Climate Change in the EU and US After the Paris Agreement.* Basingstoke [u.a.]: Palgrave Macmillan.

Wendler, Frank. 2023. "The European Green Deal Agenda After the Attack on Ukraine: Exogenous Shock Meets Policy-Making Stability." *Politics and Governance* Special Issue: Governing the EU Polycrisis: Institutional Change After the Pandemic and the War in Ukraine 11(4). doi:10.17645/pag.v11i4.7343.

Zeitlin, Jonathan, Francesco Nicoli, and Brigid Laffan. 2019. "Introduction: The European Union Beyond the Polycrisis? Integration and Politicization in an Age of Shifting Cleavages." *Journal of European Public Policy* 26(7): 963–76. doi:10.1080/13501763.2019.1619803.

4 United States

Re-launching climate action through green industrial policy

Two years into the Biden presidency, US Congress passed two major bills that proclaimed to improve infrastructure and combat inflation and are widely perceived as signature achievements of the current administration (US Congress 2021a,b, 2022a,b). They are also considered as a major breakthrough by the US towards more stringent action against climate change, and an expression of its renewed commitment to reduce GHG emissions towards net-zero by 2050 as pledged in its Nationally Determined Contribution under the Paris Agreement (UNFCCC 2020). Regardless of their specific evaluation with regard to the reduction of GHG emissions, both pieces of legislation create a new approach to pursue the goal of decarbonization through positive incentives and investment in technologies and infrastructure rather than restrictive regulation on emissions. Compared to previous stages of climate policy-making in the US (Atkinson 2018, Bailey 2015, Brewer 2015, Sussman & Daynes 2013), they also establish a new framework of institutions and governance processes for enacting action against climate change (Bang 2021) and represent a decisive departure from policies pursued by the previous Trump administration (Guliyev 2020, Glicksman 2017, Thompson et al. 2020, Kramer 2020, Leggett 2019, Danish 2018, Mehling 2017, Mehling & Vihma 2017, Selby 2019, Jotzo et al. 2018). The institutional and policy-related changes brought about by these two pieces of legislation are in the focus of this chapter.

Even more clearly than in the EU, the case of the US represents a re-orientation of climate policy towards an incentive-driven and market-based approach: combining public spending through tax breaks and direct subsidies with incentives for public and private actors to

DOI: 10.4324/9781003452041-4

invest in carbon-neutral technology and infrastructure, the BIL and IRA particularly stand out as initiatives of green industrial policy to promote carbon neutrality in conjunction with economic growth and competitiveness (CRS 2022a,b). Even though this approach to climate action is not unprecedented in the US (MacNeil 2017), its scope and size goes beyond previous programs in this regard (Larsen et al. 2022). While many observations in the previous case study have been evaluated as an aspect of policy stability, the subsequent analysis will show that key aspects of developments in the US suggest relatively stronger dynamics of disruptive change. These include the following three points.

First, the package of policies adopted to move the US towards a reduction of GHG emissions does not follow from an integrated agenda with a clearly communicated concept of climate action. As a result of the intense negotiation surrounding the adoption of legislation to enact the broader "Build Back Better" (BBB) program during the first two years of the Biden presidency, a mix of agendas and policy images has emerged that combines policy-making goals from a range of fields with the ambition of acting against climate change. The significance particularly of the IRA for progress towards decarbonizing the US economy notwithstanding, the result is an ambiguous and contested set of policy images to define the current approach of US climate action in a setting where the topic remains strongly contested both within US Congress (Guber 2021, Vandeweerdt et al. 2016) and along partisan lines within the wider public (Marlon et al. 2023). As the subsequent analysis will show, this effect of reducing the priority of climate change as an issue is created by avoiding a primary focus on the negative threat of climate change in favor of positive emphases on growth, prosperity, justice and competitiveness.

Second, the governance framework and policy process prompted by the BIL and IRA is lopsided through its almost exclusive reliance on incentives and relatively weak foundation of policies in a regulatory framework for pricing and restricting carbon emissions. This creates a notable departure from previous approaches pursued particularly during the Obama administration (Atkinson 2018, Freeman 2011, 2013). Here, initiatives for acting against climate change were primarily targeted at a (failed) attempt to establish a cap-and-trade system at the federal level and promote related initiatives for carbon pricing at the federal and state levels (Rabe 2011, 2018, Karapin 2016), and on setting restrictive emissions and efficiency standards for the energy and mobility sector (Carlson & Burtraw 2019). Instead of building on a trajectory of previously established policies, the package of policies included in the BIL/IRA package has created an almost entirely new set of decision-making venues and governance processes and a notable shift of decision-making authority to the central executive and departments in charge of the treasury, energy, transport and agriculture. As a result, policy-making under the BIL and IRA has prompted disruptive change for the set of institutions and agents of the current US climate governance framework, expanding it beyond its previous primary identification with environmental policy agents and institutions.

Finally, while a substantial impact particularly of the IRA on the reduction of GHG emissions can be assumed (Jiang 2022), a closer look reveals the degree to which the effects of the bill are uncertain and left to market processes, with regard to both the scope and the direction of change. As extant studies seeking to model the future impacts of the IRA indicate, this point applies not just to the deployment of energies and technology and resulting reductions in GHG emissions (Bistline et al. 2023, Jenkins et al. 2022, King et al. 2023) but also to the anticipated public expense through the bill and its impacts on questions of economic growth, job creation and inflation (Roy et al. 2022). In several important respects, the open-ended effects of the IRA are intended by design, especially with regard to the choice of technologies for clean energy transition and size of public expenditure through tax breaks.

These initial observations refer to the three aspects of agenda-setting, institutional venues and policy-making outcomes and are therefore closely related to the main components of our theoretical framework. As all three components of policy-making are embedded in an environment of intense political contestation (Fiorino 2022, Mann 2021, Dunlap et al. 2016, Dunlap 2019, Fisher et al. 2013, Vezirgiannidou 2013, McCright & Dunlap 2011) and denial of climate change (Collomb 2014) in the case of the US, the present chapter reconstructs their evolution with regard to the adoption of the BIL and IRA as far as they are relevant for the field of energy and climate policy. As in the previous case study, the analysis proceeds in three steps: from the reconstruction of relevant policy images to policy venues and and finally, to the evaluation of policy results concerning carbon emissions and related policy-making targets. Following on the analysis of policy images, venues and feedback (4.1–4.3), all relevant observations are again summarized in a subsequent discussion (4.4) and evaluated in a conclusion (4.5).

4.1 Policy images: from responding to the climate crisis to investing in America

As in the previous case study, the present analysis starts by evaluating the political attention assigned to climate change as a political issue at the macropolitical level of the US executive: how much emphasis was laid on this issue by President Biden and other key members of his administration as a political problem? A first indication is provided by a description of the "immediate priorities" of the incoming Biden–Harris administration published on the White House website: here, the issue of "climate" is listed as the second of overall seven priorities, promising that "President Biden will take swift action to tackle the climate crisis", "mobilizing a whole-government effort", and "put the United States on a path to achieve net-zero emissions, economy-wide, by no later than 2050" (White House 2024). This pledge marks a departure from the approach of the previous Trump administration and raises the profile of climate change as an important issue in the agenda of the White House. But how far can it be traced as a political topic of high importance through the agenda-setting and policy-making actions of the Biden administration since its arrival in office?

A first general survey of emphases on the topic of climate change over time is gained from the analysis of the Inaugural and subsequent State of the Union Addresses delivered by President Biden between the years 2021 and 2024. These addresses present the broadest thematic outlook on political issues considered as important by the White House at the highest macropolitical level. It is important to add that the survey below does not seek to present a complete mapping of all political topics covered in these speeches; instead, it presents data on the relative frequency of references in these speeches to the issue of climate change and the environment directly, and issues brought in relation with them in subsequent legislation: namely, economic recovery and prosperity, social security and justice, and questions of geopolitical security. As in the case of the EU, these references were identified using automated coding based on a dictionary compiled from word frequency rankings and subsequent word-in-context analysis. The results are shown in Figure 4.1 below, listing both keyword

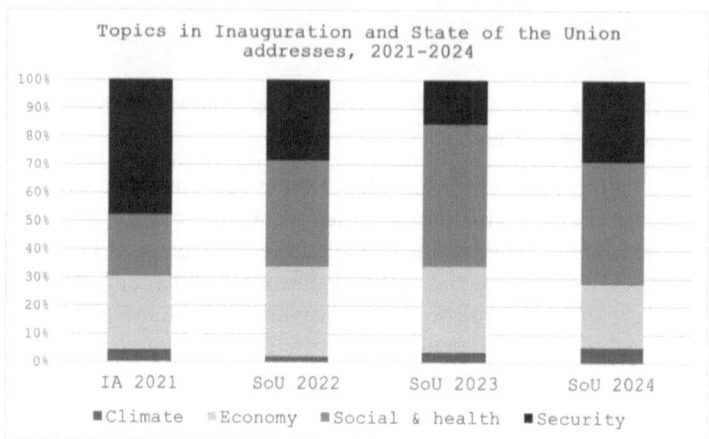

Figure 4.1 Topics in Inauguration and State of the Union addresses

counts in absolute numbers and their relative frequency in a comparison between thematic segments associated with those keywords.

While this survey confirms regular references by President Biden to the issue of climate change (mentioned 11 times and including six references to the term "climate crisis"), it also puts the relative emphasis given to this topic in perspective. Other political issues – primarily those related to jobs, workers and living standards, the pandemic and health issues – dominate in comparison to the scarcer references to environmental and climate issues. Interestingly, references to Russia, Ukraine and Russian president Vladimir Putin do not spike as a consequence of the Russian attack in 2022 as in the EU case, with the emphasis of the State of the Union Addresses laid primarily on the level of domestic US politics and policy-making.

From this point of departure, a second step is to inquire to what degree President Biden and other key members of his administration have referred to climate change in their public communication about the two key legislative acts covered in this chapter. Covering a broad range of issues ranging from inflation control to jobs and healthcare, the BIL and IRA leave discretion to political agents how strongly they refer to them as bills adopted to reduce carbon emissions, mitigate climate change and address other environmental issues. In order to gauge the relative attention to climate in relation to other issues,

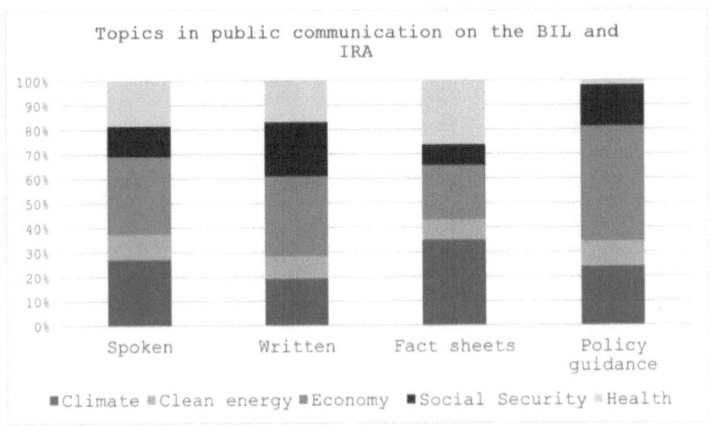

Figure 4.2 Topics in public communication on the Inflation Reduction Act and Bipartisan Infrastructure Law, 2022–3

the survey below presents data on thematic references in the public communication of the White House. These include three sets of documents: first, transcripts of spoken statements by President Biden and Vice President Harris at occasions such as the signing of the IRA, its anniversary and related events; second, written statements issued by the White House on key targets and priorities of the IRA and BIL; and finally, introductory parts of guidance documents on both acts of legislation (White House 2022, 2023). Comprising a corpus of some 38,000 words, this set of documents is considered as a key component of political communication by the White House on its agenda and therefore what priority is assigned to the issue of climate as a component of the BIL and IRA (Figure 4.2).

This analysis confirms a significant proportion of references to climate and environmental issues in the public communications of President Biden and Vice President Harris on the IRA; however, it is also evident that its rationales and promised effects reach across a broader range of policies, including aspects of economic growth, social security and healthcare. The fact sheets surveyed also cover the provisions of the IRA in its entirety, whereas the excerpts from policy guidance documents cover those publications focusing on the clean energy and climate aspects of the IRA. Even here, however,

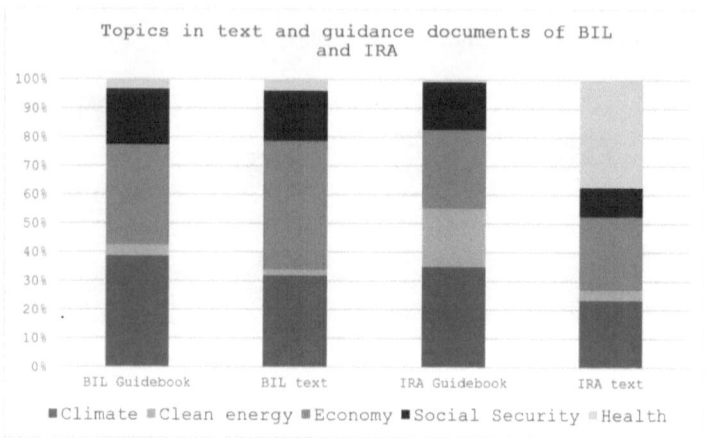

Figure 4.3 Topics in legislative text of Bipartisan Infrastructure Law and Inflation Reduction Act and guidance documents

references to economic terms such as jobs, businesses, manufacturing or investment outweigh those referring to effects of climate change or terms associated with a transition to clean energy.

To conclude this survey, our last step is to apply the automated keyword analysis to the legislation itself: to what degree does the text of the IRA and BIL bills refer to terms associated with climate change? The graph below (Figure 4.3) presents a keyword survey for the text of the BIL and IRA together with the detailed guidance documents issued by the White House.

This overview leads to substantial insights about both the salience of climate change and its linkage to other related topics. While the re-launch of climate policy by the Biden administration stops short of achieving top salience within agenda-setting communication at the highest political level, it is the subject of a substantive policy discourse by the White House and relevant departments in relation to two of the signature legislative achievements of this administration. The very broad range of topics included in this policy discourse includes references not just to economic and energy issues but also to social and health issues, as demonstrated particularly in the survey of thematic references in fact sheets, guidance documents and the text particularly of the IRA. These observations lead to the main question of this section: namely, how to define the policy images of climate action pursued by the Biden administration from its initial announcements of action through the legislative negotiation of the BIL and IRA. In a more in-depth, qualitative review of relevant documents, the policy image at the core of current US climate action presents itself as an evolving set of policy ideas that relate and connect a broad range of policies and policy-making priorities. Concerning the issue of climate change, the dominant policy image varies considerably in a sequence of decision-making actions, from the initial executive actions of the Biden administration to the complex and contested legislative negotiations dealing with how to integrate climate-related targets into a range of economic and industrial policies.

Contrary to the EU case, no comprehensive policy-making agenda laid down in a unified document such as the EGD communication exists that could be reviewed as the framework for climate action across these different stages of decision-making. Instead, several points of reference need to be considered, including executive acts and statements by the White House, guidance documents by key departments and the Environmental Protection Agency (EPA) covering relevant legislation, and of course policy programs adopted particularly within the BIL and IRA. The present analysis traces the evolution of the climate policy image through three relevant stages, from its initial definition at the executive level through its integration into legislative acts to subsequent efforts by the Biden administration to communicate decarbonization as part of its overall agenda.

As a point of departure, the clearest account of the climate policy ambitions pursued by the Biden administration is the Executive Order "Tackling the Climate Crisis at Home and Abroad" (EO 14008), signed by President Biden on January 27, 2021 and followed by Executive Orders on environmental justice, local communities and the implementation of the IRA (Federal Register 2021a,b, 2022a,b, 2023, ClimateWire 2022). The first of these orders is particularly remarkable in the clarity with which it establishes two principles for the approach of the Biden administration towards climate change: namely, its centrality for all fields of foreign and security policy, and the adoption of a government-wide approach to the climate crisis involving all relevant departments and agencies.

In the opening sections of the first of its two main parts, the order states that the US and the world face a "profound climate crisis" that only leaves a "narrow moment to pursue action" in order to "avoid setting the world on a dangerous, potentially catastrophic climate trajectory" (Federal Register 2021: 7619). The perception of climate change as a crisis is frequently repeated in subsequent sections, stating that it "threatens our people and communities [...] and starkly, our ability to live on planet Earth" (ibid.: 7622). Based on several references to the "security implications of climate change" (ibid.: 7621), the order concludes that climate considerations must be an essential element of US foreign policy and national security. From this point of departure, the second main part proclaims a government-wide approach that must use "bold, progressive action" to "achieve net-zero emissions, economy-wide, by no later than 2050" and be based on the principle of environmental justice (ibid.: 7622). As additional interim targets, the order spells out the ambition to create a carbon pollution-free energy grid no later than 2035, doubling off-shore wind energy and conserving at least 30% of lands and waters by 2030, and to eliminate fossil fuel subsidies in the federal budget from 2022 onwards. Concerning the institutional framework for the enactment of climate goals, the main significance of the order is to create the Climate Policy Office headed by a National Climate Advisor within the White House as the central hub for coordinating policies. Furthermore, a National Climate Task Force is created convening all relevant departments and agencies of the executive branch (including Treasury, Defense, Interior, Agriculture, Commerce, Housing and Labor) to deliver the government-wide approach to climate action proclaimed at the outset.

Creating a clear and ambitious policy image, the Executive Order sets out the policy framework envisaged for climate action by the Biden administration, based on its commitment to the Justice40 initiative as a cross-cutting principle to direct 40% of investment under specified programs to disadvantaged communities as an expression of environmental justice (Perls 2022). This principle is further laid out in more detail in the previous Executive Order 13990 of January 20, 2021, which makes only passing mention of climate change but discusses a range of policy issues relevant in this regard. Among these is the withdrawal of permits for drilling in the Arctic Refuge and for building the Keystone XL pipeline, proposing new standards for the reduction of methane emissions and re-launching a strategy for assessing the social cost of carbon emissions (Federal Register 2021b: 7038–40). The triad of references to climate change as crisis, the adoption of a government-wide approach and environmental justice creates a policy image that covers the discursive framing, institutional framework and policy approach of future climate action. While not being legally binding for policy-making at the federal level beyond the creation of government units and offices, the order is relevant for defining the framework of priorities and policies proposed by the Biden administration.

This relatively stringent and coherent policy image proclaimed in Executive Orders has, however, been rendered more complex and multi-faceted through the transposition of proposed policy goals into applicable legislation during the years 2021 and 2022. Tracing the policy image of climate action through the stages of legislation from the initial BBB proposal to the adoption of the IRA shows two major modifications: first, a shift towards de-emphasizing the salience and priority assigned to the climate crisis in relation to priorities of non-environmental policy-making fields; and second, an increasing complexity of messages and corresponding loss in the clarity and consistency of the emotive appeals and empirical statements used to define a policy image for US climate action.

The main gateway for the transposition of proclaimed climate action targets into legislation is the BBB agenda promoted by President Biden since the launch of his campaign for office. As a proposed policy program, the agenda creates a framework of policies that were never realized in a single comprehensive bill but gradually enacted with major modifications and reductions through several pieces of legislation. A first major legislative action resulting from the BBB agenda and shaping parts of its subsequent policy image was the adoption of the BIL (officially termed the Infrastructure Investment and Jobs Act, H.R. 3684 and signed into law after passing both chambers of Congress on November 15, 2021). Several of the bill's components are considered to be relevant as a step towards more stringent action towards both the mitigation of and adaptation to climate change; among these are provisions for investment in the upgrading of power grids and more energy-efficient technologies in public and private buildings, support for investment in technologies related to the energy transition such as batteries, and programs to strengthen the resilience of communities against extreme weather events. The overall policy image of the BIL, however, is defined as a broad investment in the modernization of public assets and services covering primarily the fields of transportation, broadband, energy grids and water infrastructure including pipes, water storage and wastewater treatment.

The major legislative breakthrough to result from the negotiation of the BBB agenda in Congress is the IRA (H.R. 5376, signed into law on August 16, 2022 after passing the House initially in November 2021 and clearing the Senate on August 7, 2022). Praised as the most relevant breakthrough piece of legislation for climate action in US history, the policy image at the core of the bill is a complex and confusing one, evidently addressing climate change only in conjunction with and even as a secondary effect of other, more foregrounded priorities. Starting with its title as a bill aiming to reduce inflation, the IRA considers decarbonization as an effect of investment and deployment of profitable technologies but not as a defining principle; the mitigation of climate change is effectively submerged as a secondary effect under more clearly prioritized targets of modifying taxation, addressing the affordability of healthcare, and setting incentives for promoting clean energy. The fact that none of its opening clauses or major chapters directly address the issue of climate stands in notable contrast to the fact that concerning its content, the IRA creates the most relevant legislative breakthrough for climate policy at the federal level in the US.

This shift of the policy image identified with the IRA is caused by three main factors related to the political and institutional environment of its passage: first, the decision to promote decarbonization targets in conjunction with a wider range of fiscal, economic, geopolitical and justice-related targets to secure political support from currents of the Democratic party as opposed to a regulatory intervention into market freedoms; second, the requirement to pass IRA as a budget reconciliation bill to avoid filibuster in the US Senate and be able to pass the act using the tie-breaking vote of Vice President Harris in the 50–50 split chamber, with the effect that the main contents of the legislation had to be focused on spending rather than prescription of policy; and finally and most evidently, the accommodation of positions in the US Senate against the initial "BBB" bill and felt primarily through the reduction of spending programs and clearer inclusion of provisions to protect producer interests within specific constituencies, targeted particularly through the inclusion of domestic content requirements for tax breaks and subsidies.

Defining the policy image of the IRA is challenging, considering its scope and complexity as a piece of legislation covering issues of taxation, healthcare and investment in a broad range of technologies and infrastructure. Indeed, the most fitting characterization of the

IRA's policy image would be that it lacks one, as a broad variety of agenda-setting points and messages are identified with its passage as a piece of legislation with macroeconomic, geopolitical, industrial and social policy aspects. Concerning the climate and energy-related provisions, however, the clearest such image that can be distilled comes from guidance documents issued by the White House on the content and main intentions of the IRA. In its opening paragraphs, action on climate change is presented as an "opportunity to lower costs for all Americans, create good-paying union jobs for workers, address the cumulative impacts of pollution on disadvantaged communities, and ensure America leads the global clean energy economy" (White House 2023: 3). While presenting the act as an expression of American leadership in confronting the "existential threat of the climate crisis" (ibid.: 5), the subsequent outline of the legislation's content amalgamates four main policy targets: the prosperity and social security of middle-class families; targeted support for underprivileged or deindustrialized communities and minorities based on the idea of environmental justice; supporting economic growth through the lowering of energy bills and job creation through the provision of opportunities for investment; and securing the technological modernization and competitiveness of the US economy in sectors related to the clean energy transition in a global context. From this point of departure, the most relevant climate-related pledge of the IRA is to achieve a transition towards a 40% reduction of GHG emissions relative to 2005 by 2030, stopping short of but contributing significantly to the commitment of the Biden administration to reduce GHGs by at least 50% overall and achieve carbon neutrality of the energy grid by 2035.

At the core of the policy image proposed for the IRA, therefore, is the idea of synergy between social and economic policy goals and the mitigation of climate change, whose urgency is referred to but not laid out in the sense of a necessary constraint or limitation for future growth or consumption. The political framing of climate action by the Biden administration is very consequent in referring to GHG emission reductions as a positive side-effect of economic and technological opportunity and recovery; any negative references to the threatening aspects of the climate crisis and resulting steps towards restraint, regulation or limitation are avoided. In this sense, it is intriguing how the IRA has a dual presence as both the most significant piece of climate legislation at the US federal level and an act that includes only intermittent references to the problem of climate change or GHG emission reductions.

This aspect is also present in the public communication by the Biden administration of the enactment of the IRA. To summarize, the first step of our analysis demonstrates the contrast between the seriousness of ambitions within the Biden administration to move towards carbon neutrality and the dispersed and submerged climate policy image resulting from the transposition of these ambitions into relevant legislation. The table below (Table 4.1) presents a survey of the main emotive appeals and acts of empirical information underlying relevant Executive Orders as the point of departure for climate action

Table 4.1 Policy images of US climate action during the Biden administration

	Emotive appeals	*Empirical information*
Executive Order "Climate Crisis at home and abroad" (EO 14008, January 21, 2021)	(1) Climate crisis defines US foreign policy and home security (2) Adopting a government-wide approach to climate action	(1) US will re-engage in global effort to combat climate change (2) US commits to path towards carbon neutrality by 2050
Infrastructure Act (BIL)	(1) Historic investment in US infrastructure and competitiveness (2) Create opportunities and fairness, leave no one behind	(1) Prioritized funding for transportation and clean energy (2) Boost manufacturing in US, improve poor communities
Inflation Reduction Act	(1) Support middle-class jobs, incomes and prosperity (2) Control inflation by taxing corporations / bond purchases (3) Defeat special interests and reduce the federal deficit (4) Apply environmental justice and curb global warming	(1) Lower healthcare and energy costs, create millions of jobs (2) Set 15% minimum corporate tax and raise $124bn in revenue (3) Lower prescription drug costs and improve labor standards (4) Target pollution and move US towards carbon neutrality

Sources: CRS 2022a, 2022b, EPA 2022a, 2022b, Evergreen 2022, US Congress 2021a, 2022a, White House 2023.

by the Biden administration and its main legislative achievements and subsequent public communication. Considering the scope and ambition of the policies covered here, it is clear that both the content and policy image particularly of the IRA promotes major policy change and implies innovation with regard to the venues and governance framework of climate action, covered in the next section.

4.2 Policy venues: unpacking the governance framework of US green industrial policy

The enactment of the IRA has had a disruptive impact on policy venues and subsystems relevant for US climate action, particularly as a vector of expansion and hierarchical direction in comparison to previous stages. Reconstructions of the evolution of US climate governance before the Biden administration have highlighted its weak and volatile institutional foundation (Brewer 2015, Atkinson 2018, Karapin 2016); during these stages, initiatives to promote a reduction of GHG emissions were almost exclusively enacted through delegation of standard-setting to the EPA. In addition, the involvement of other departments of the executive branch has been limited mostly to the appraisal of climate-related risks and subsequent action, particularly those in charge of homeland security and defense. The absence of a comprehensive policy-making framework for climate action at the federal level at these stages is particularly due to political resistance against legislation establishing regulation on key issues such as carbon pricing within Congress and a strong culture of litigation between agents at the state level against the federal level (Thompson 2020, Sussman & Daynes 2013). In comparison, the launch of a green industrial policy particularly through the IRA has reinforced the governance function of established institutions such as the EPA; beyond that, it has also created a new and still evolving framework of policy-making that involves a range of departments and requires strong coordination at the level of the central executive. The structure of this governance framework is prescribed by reference to responsible departments within the eight titles of the IRA (cp. CRS 2022).

The systematization of policies comprising the IRA can be approached in different ways, including: their foundation in one of the eight titles and related Congressional Committees covering finance, commerce, agriculture or energy; the type of policy and recipient in a distinction between tax incentives, provision of loans and grants, regulation or technical guidance; the different sectors covered, such as industry, buildings, energy or transportation, and their relevance for curbing GHG emissions; or the relative size of available resources and funding. The approach taken here reviews the IRA with regard to the emerging climate governance framework of the US and its key policy-making agents, venues and forms of coordination and political direction. Accordingly, the subsequent review focuses its systematization on the degree of involvement and mutual coordination of departments and agencies under the direction of the central executive in the White House. This survey is summarized in Table 4.2 below.

As this overview demonstrates, the IRA is a vast piece of legislation whose implementation is not to be understood as a simple process of top-down enactment. Instead, it is more appropriately defined as the evolution of a new governance framework that involves a range of departments and agencies; its implementation depends on a range of public and private stakeholders and companies at least to the same degree as on activity by the US federal government. Within this framework, three nodal points particularly stand out as the most relevant.

First, the implementation of investment programs particularly under the IRA puts the Department of Treasury (DoT) firmly at the center of the new climate governance framework in the US. It is here that the some of the most substantial programs for incentivizing investment in clean energies and mobility – particularly the investment and production tax credits (ITC and PTC) for renewable energy and subsidies for zero-emission cars that can be considered the main core of the IRA's policy content – are administered in cooperation with the Internal Revenue Service. In this context, the creation of a Climate Hub in the Office of the Secretary establishes a new nodal point for coordinating efforts and information flows in the implementation of US domestic climate policy. Deputy Secretary of the Treasury Wally Adeyemo, as the chief operating officer in charge of the administration's economic recovery strategy, has emerged as one of the primary agents for enacting and communicating steps taken towards the implementation of the IRA agenda, in conjunction with senior officials in charge of tax policy such as Assistant Secretary Lily Batchelder.

A second pillar for enacting the IRA is the Department of Energy (DoE), whose role focuses on some of the key supply-side innovations introduced with regard to the innovating of grids, permit procedures, and promotion of clean technology production and development. In

Table 4.2 Key climate and energy policy programs and departments within the Inflation Reduction Act

	Policy goals and recipients	Resources and instruments
DoE	(1) Promotion of renewable / **clean energy** technologies, including wind, solar, CCS, nuclear and critical minerals processing	(1) $40bn loan authority administered through LPO supported by $3.6bn in credit subsidy
	(2) Retool, repurpose or replace energy **infrastructure** and enable operating plants to sequester GHGs	(2) Up to $250bn loan authority to support projects, supported by $5bn in credit subsidy for DoE
	(3) Construction or modification of electric **grids** designated in national interest	(3) $2bn in funding for direct loan program, available through 2030
	(4) Support and accelerate **siting / permitting** of inter-state transmission lines	(4) $760mn in grants provided to siting authority or state, local entities
	(5) Support domestic production of clean **energy cars** and their components	(5) $3bn for loans to manufacturers, $2bn for retooling production lines
	(6) Support **emissions-intensive industries** sectors to install GHG-reducing technology	(6) $5.8bn in support for iron, steel, cement glass, paper, alu industries
	(7) Support homeowners, states and Tribes to improve **energy efficiency of homes**	(7) Close to $9bn in funds for rebate programs (heat pumps, insulation etc.)
DoT	(1) Support **renewable energy / clean energy** production and investment (two stages, renewables pre-2025 and clean energy after)	(1) Extension of extant tax credits, technology neutral, emission-based ITCs and PTCs from 2025
	(2) Support domestic manufacturing of clean energy **technologies** (solar, wind, battery cells, critical mineral processing)	(2) Production credit (uncapped), permanent for critical minerals and available 2023–32 for other items

(Continued)

Table 4.2 (Continued)

	Policy goals and recipients	Resources and instruments
	(3) Support production or recycling of **renewable** and energy-**efficiency equipment** and CCS	(3) Authority to allocate $10bn to eligible projects, with requirement to direct 40% to energy communities
	(4) Subsidize purchase of qualifying new, used and commercial **clean vehicles**	(4) Uncapped subsidy, up to $7500 for new and $4000 for used cars
	(5) Support installation of **residential clean energy** (rooftop solar) and efficiency measures	(5) Provision of 30% tax credit to homeowners, additional tax credits
DoA	(1) Support rural electric cooperatives to improve energy efficiency and deploy clean energy / CCS systems	(1) $9.7bn for loan and guarantee programs for eligible rural cooperatives
	(2) Support agricultural producers and rural areas to construct **renewable** energy systems	(2) $2bn loan financing and grants (REAP); $1bn in loans for rural areas
	(3) Provide technical and financial assistance for producers (**soil carbon, sequestering**)	(3) $8.45bn in funds in the Environmental Quality Incentives Program
	(4) Support conservation projects to help producers to **reduce and sequester GHGs**	(4) $4.95bn in the Regional Conservation Partnership Program
DoH	(1) Support **retrofitting** of HUD-assisted properties, state governments and businesses	(1) $2bn in support for funding for owners and governments
EPA	(1) GHG Reduction Fund for clean energy projects in **disadvantaged communities**	(1) $27bn for awarding grants to mobilize and leverage private capital
	(2) Assist community organizations to reduce **air pollution** and improve climate resilience	(2) $3bn in block grants and technical assistance for eligible organizations
	(3) Support **Tribes, states and local** governments to reduce GHG emissions	(3) $5bn in grants to develop and implement grants for eligible entities

Multi-agency	Strengthen and accelerate **federal permitting** and environmental reviews (grids, projects)	Over $1bn in support for Steering Council and Council on Environmental Quality
Cross-cutting	(1) Support **state, local and Tribal governments** not eligible for income tax credits	Eligibility for direct pay and transferability for major IRA programs

CCS = carbon capture and storage; HUD = Department for Housing and Urban Development.

Overview of provisions by departments:

DoE: (1) = Funding for DoE LPO, IRA Section 50141; (2) = Energy Infrastructure Reinvestment Financing, IRA Section 50144; (3) = Transmission Facility Financing, IRA Section 50151; (4) = Grants to Facilitate Siting of Interstate Electricity Transmission Lines, IRA Section 50152; (5) = Advanced Technology Vehicle Manufacturing Loan Program, IRA Section 50161; (6) = Advanced Industrial Facilities Deployment Program, IRA Sections 50142 and 50143; (7) = Home Energy Performance-Based, Whole House Rebates, High-Efficiency Electric Home Rebate Program, State-based Home Efficiency Contractor Training Grants, IRA Sections 50121, 50122 and 50123.

DoT: (1) PTC for Electricity from Renewables; ITC for Energy Property; Clean Electricity PTC; Clean Electricity ITC; IRA Sections 13101, 13102, 13701, 13702(h); (2) = Advanced Manufacturing Production Credit, IRA Section 13502; (3) = Extension and Expansion of the Advanced Energy Project Credit, IRA Section 13501; (4) = Clean Vehicle Credit, Previously-Owned Clean Vehicles Credit and Commercial Vehicles Credit, IRA Sections 13401, 13402 and 13403; (5) = Energy Efficient Home Improvement Credit, Residential Clean Energy Credit, New Energy Efficient Homes Credit, IRA Sections 13301, 13302, 13304.

DoA: (1) = USDA Assistance for Rural Electric Cooperatives, IRA Section 22004; (2) = Rural Energy for America Program (REAP) and Electric Loans for Renewable Energy, IRA Sections 2202 (a,b) and 22001; (3) = Environmental Quality Incentives Program, IRA Section 21001 (a) (1); (4) = Regional Conservation Partnership Program, IRA Section 21001 (a).

DoH: (1) = Green and Resilient Retrofit Program, Assistance for Latest and Zero Building Energy Code Adoption, IRA Sections 30002 (a) (1,3,4) and 50131.

EPA: (1) = Greenhouse Gas Reduction Fund, IRA Section 60103; (2) = Environmental and Climate Justice Block Grants, IRA Section 60201.

Multi-agency: Funding for Departments of Agriculture, Energy, Interior, Transportation, the EPA, the National Oceanic and Atmospheric Administration, the US Forest Service, the Federal Energy Regulatory Commission, the EPA, Federal Highway Administration, Federal Permitting Improvement Steering Council and Council on Environmental Quality; IRA Sections 23001, 40003, 50301, 50302, 50303, 60115, 60402, 60505 and 70007.

Cross-cutting: Advanced Energy Project Credit, IRA Section 13501, Tax Code Section 48C.

Sources: CRS 2022a, 2022b, DoE 2023, EPA 2022b, Federal Register 2022b, US Congress 2022a, White House 2023.

terms of practical enactment, the Loan Programs Office (LPO) within the DoE is of primary importance as the department's in-house bank whose volume and lending power has been substantially expanded by passage of the IRA.

Finally, the EPA continues to have an important function in the development of US energy and climate policy through both its established and several new roles and instruments. This role involves both distributive and regulatory instruments. One major component of the IRA for directing funds to finance projects to reduce GHGs is the Greenhouse Gas Reduction Fund (ClimateWire 2024e,m). This fund is administered by the EPA and appropriating $27bn in funding based on provisions of the Clean Air Act as amended by the IRA and provided through competitive grants under the heading of three main programs (namely, the National Clean Investment Fund, the Clean Communities Investment Accelerator and the Solar for All program; cp. ClimateWire 2024e,m). At the level of regulation, the IRA is significant for the activity of EPA in two further respects: first, by defining GHGs including carbon dioxide and methane as pollutants, it provides a new legal basis for the EPA to issue rules for the limitation of emissions from the power sector, resulting in the issuing in April 2024 of rules for the phase-out of fossil fuel plants by 2039 or introduction of carbon capture mechanisms; and second, by mandating the EPA to revise extant rules for the reporting of methane emissions, particularly for oil and gas extraction, and introducing a new methane fee to be raised from 2025 as effectively the first form of GHG taxing in the history of US climate policy (ClimateWire 2024a,d). First details of revised reporting mechanisms and the application and exemptions of the fee were announced by the EPA in January 2024.

While the three parts of the executive branch highlighted here – namely, the DoT, DoE and EPA – create the primary pillars for the IRA governance framework, additional departments are also involved, including with programs that comprise billions of dollars in public spending. In this regard, the Department of Agriculture (DoA) with its role of supporting producers to adopt green technologies and promote carbon sequestration, and the Department of Housing (DoH) with its authority over publicly owned properties, are especially notable.

The policy-making dynamic prompted by the IRA creates considerable challenges of coordination and monitoring; these challenges arise from both the range of departments and policy fields involved and the task of ensuring compliance with assigned criteria of social and environmental conditionality attached to the disbursal of very large sums of public subsidies and tax credits. To ensure this coordination, an institutional framework for the direction of governance processes under the IRA has been created through the establishment of an Office on Clean Energy Innovation and Implementation in the White House by Executive Order 14082 of September 2022 on the Implementation of the Energy and Infrastructure Provisions of the IRA (Federal Register 2022). The Office is headed by John Podesta who in this function received the title of Senior Advisor for Clean Energy to the President and also acts as the chairman of the National Climate Task Force that convenes the leadership of all relevant executive departments and relevant entities such as the EPA and the Office of Management and Budget, and also includes National Climate Advisor Ali Zaidi as vice-chair. Through this institutional framework, a structure for the vertical direction of policies launched under the IRA has been created that creates a link between the Executive Office of the White House and relevant departments, and establishes a connected but separate governance framework in relation to other aspects of domestic climate policy headed by the National Climate Advisor.

Podesta is widely seen as an influential figure for enacting the IRA provisions, given his seniority as a campaigner and policy-maker for previous Democratic administrations and his involvement in the negotiation of the Paris Accord in 2015. Most recently, Podesta's role for directing climate action by the Biden administration has been further increased as he also assumed the role of global climate policy envoy after the departure of John Kerry from this post in early March 2024 (ClimateWire 2024f,g,k). Podesta has not formally assumed Kerry's formal title to avoid a formal Senate confirmation procedure for being assigned to this post because of expected pushback in the chamber (ClimateWire 2024c). In political terms, however, he is now de facto directing teams for the domestic clean energy and global aspects of climate action located at the White House and the State Department, thereby strengthening centralized leadership of all aspects of US climate policy from the White House.

In addition to these aspects of vertical direction, the launch of the IRA also brings innovation in the creation of structures for horizontal coordination. Concerning the cross-cutting issue of environmental justice, an important aspect is the Office of Energy Justice and Equity (renamed in October 2023 from the previous title Office for Economic Impact and Diversity) in the DoE, led by its director Shalanda Baker.

Beneath these levels of political leadership, an entire set of task forces and inter-agency groups has been created to ensure coordination between the various departments and agencies involved in the enactment of the IRA. The Climate Task Force established by EO 14008 and comprising virtually all departments of the executive branch is only the most senior of these coordinating venues. In addition, several inter-agency groups with more thematically specific tasks were created to ensure coordination, as listed in the Table 4.3 below.

To summarize, the present analysis shows how the launch of investment policies particularly under the IRA has not just established a broader policy image for US climate action but also prompted the creation of a governance framework in its own right. It considerably exceeds the previous framework of energy and climate policy in terms of its scope and competences, and assigns a key role particularly to the DoT and DoE beyond the established lead function of the EPA. The result is the establishment of new policy subsystems under the broad headings of energy, mobility, buildings and land use that are operated in different administrative settings but linked through a set of cross-cutting principles. These include, most importantly, the Justice40 principle and support for non-taxable entities through the instrument of direct pay, and are part of a broader political framework of coordination and leadership from the Executive Office of the White House. Put more succinctly, the re-launch of climate action through the BIL and IRA is not just a process of policy-making but also prompts institutional change within the executive branch of the US political system.

4.3 Policy feedback: the Inflation Reduction Act as a framework for dynamic, open-ended policy change

In this third stage of our analysis, we turn to the assessment of policy-making results in terms of their effects for action against climate change and related policy goals discussed in the previous sections. It is evident that the extent of policy change prompted particularly by the IRA cannot be fully evaluated only two years after its adoption and considering that many of its provisions reach into the period after the next US presidential election and even the next decade. In fact, many of the most relevant tax breaks of the IRA are set to expire only in 2036. Taking this into account, however, a sufficiently large range of studies have emerged to date to model policy-making trajectories

Table 4.3 Political leadership and coordination mechanisms for the Inflation Reduction Act governance framework

	Relevant units / agents	Functions of coordination and oversight
Central executive / White House	(1) National Climate Advisor (2) Advisor for Clean Energy Innovation and Implementation (3) Climate Policy Office (4) National Climate Task Force	(1) est. by EO 14008 (2) est. by EO 14082 (3) led by National Climate Advisor (4) est. by EO 14008, convenes all relevant departments and agencies to organize government-wide approach, chaired by Senior Advisor
Department offices and coordination	(1) LPO at the DoE (2) Office for Environmental Justice at the Department of Health (3) Interagency Working Group on Coal and Power Plant Communities (4) Environmental Justice Interagency Council (5) Interagency Working Group to Decrease Risk of Climate Change (6) Interagency Working Group on Social Cost of Greenhouse Gases (7) Task Forces for Implementation of BIL and IRA	(1) est. previously (2) est. by EO 14008 (3) est. by EO 14008 (4) est. by EO 14008 (5) est. by EO 14008 (6) (Re-)established by EO 13990 (7) Established by EOs 14052 and 14082
EPA	(1) Implementing function (2) Regulatory function: standard-setting and application (3) Oversight and advisory function	(1) GHG reduction fund (2) GHG emission standards (energy, vehicles, methane) (3) Methane reporting and fees

Sources: CRS 2022a, EPA 2022a, 2022b, Federal Register 2021a, 2021b, 2022b.

to discuss its expectable results against the backdrop of the distinction between negative and positive feedback.

The subsequent evaluation of the IRA and BIL can rely on two major sets of sources: first, reports and surveys issued by US government institutions such as the DoE and the EPA (DoE 2023, EPA 2022); and second, studies by independent think tanks such as the Rhodium Group and research networks with an academic grounding such as the REPEAT project (Jenkins et al. 2022, 2023). The tables below (Tables 4.4. and 4.5) present a survey of the main predictions of overall policy-making results, sectoral distinctions and major priorities and challenges identified for future implementation, both with regard to key climate and energy targets and with regard to a wider set of economic, social and more global (such as energy security) targets.

Across relevant policy-making fields, a key insight from extant prognoses is the degree of variation of expected effects of the IRA, reflected primarily in the distinction of more optimistic, pessimistic and mid-range trajectories. Among the key factors affecting future developments in a comparison of these scenarios are responses by consumers and companies to spending and investment incentives offered by the IRA; sources of uncertainty also include future economic developments, including factors such as growth rates, inflation and prices of required resources, as well as political factors affecting the cooperation of relevant agents at the state and federal levels. Beyond these sources of uncertainty, the studies reviewed here broadly agree on three findings concerning the climate- and energy-related goals of the IRA.

First, while all extant studies expect a considerable acceleration of progress in moving the US towards decarbonization, a significant gap remains between the proclaimed ambition of achieving carbon neutrality by mid-century and the anticipated effects of policies enacted under the BIL and IRA. As a general benchmark, GHG emissions by the US are required to decrease by an average of 6.9% between the current stage and 2030 to meet the Biden administration's pledged goal of a reduction of 50–52% by that time; while estimates of actual current reductions vary between about 2% and just above 3% for the year 2023 according to different estimates, it is therefore clear that a clear gap between ambitions and effect still exists (ClimateWire 2024j,l). One of the most comprehensive studies of the anticipated decarbonization effects of the IRA published by the Rhodium Group expects a fulfillment of the interim 2030 goal of around 50% GHG emissions relative to 2005 only in its most optimistic scenario and in combination with additional regulatory action targeting energy, transportation and agriculture. n its mid-level and more pessimistic prognosis, this reduction is expected to reach only between 29% and 42% GHG reduction by 2030 and between 32% and 51% by 2035, stopping substantially short of the targeted goal of halving emissions by 2030 (King et al. 2023: 5).

Second, extant studies agree on the asymmetric effects of current policies across sectors relevant for decarbonization, with the strongest dynamic expected in energy production and to a more limited extent in mobility, but slower effects of change with regard to buildings, industry and agriculture, as detailed in the overview below. Finally, several factors beyond the cost and economic returns of investments and technology are identified as key challenges for the future implementation of policies promoted through the BIL and IRA. Among these are obstacles against the acceleration of siting and permitting procedures for grids and infrastructure; the availability of skilled workers and administrators for enacting required efforts of investment and installation; and the availability and prices of required materials and components as affected by relevant supply chains and trade relations. Due to these factors, estimates about the expected effects of the IRA on decarbonization still generally include relatively wide margins of variation between optimistic and more critical scenarios (Table 4.4).

Beyond extant uncertainties concerning the prognosis of future GHG emission development, a deliberate component of the IRA is its open-endedness; this concerns the avoidance of a top-down regulatory intervention aiming at the promotion of certain technologies in favor of a market-based process of competition between available options involved in the decarbonization of the energy grid. This point corresponds to the general approach of the IRA to prompt policy change that is defined as subject to competitive dynamics: no general or sector-specific targets of GHG emissions reductions are specified or enacted as legally binding in any part of this legislation, thereby distinguishing it fundamentally from a climate law prescribing a legally binding commitment to decarbonization. An open-ended approach also defines some of the key features of the IRA beyond its climate- and energy-related provisions. Most important among these is the amount of envisaged overall public spending: in spite of the highly publicized announcement of $369bn as the amount available for investment in

Table 4.4 Expected impacts of the Inflation Reduction Act on energy and climate policy-making goals

	Overall expected effect on GHG emissions reduction	Differentiation by sectors	Key priorities / challenges for further action
DoE	35–41% decline of GHG emissions by 2030 relative to 2005 levels (ca. 27% without BIL / IRA). Estimated reduction of CO_2 emissions by ~ 1,000 metric tons by 2030	Greatest reduction in power sector (renewables increase to 72–81% by 2030). Transport: 19–24% CO_2 emission reduction by 2030, ZEV share up to 65% in optimistic scenario. 33–42% decrease of industry emissions of GHGs by 2030	Strengthening US energy security. Lowering energy costs for households and industry. Investment in disadvantaged and energy communities. Strengthening domestic production of clean energy technologies
Rhodium Group	29–42% by 2030, 32–51% by 2035, depending on high-, mid- or low-emissions scenario. Paris targets reached only in most optimistic scenario; additional action required	Detailed evaluation: dynamic of change in power (-45–75% by 2035) and transport (-15–32%). Greater obstacles in buildings, agriculture and carbon removal	**Core challenges / factors:** Non-cost factors for clean energy deployment (permitting, skills, supply). Rate of economic growth by 2030. Developments at state and local level
REPEAT project	37–41% GHG reductions by 2030, 46–53% by 2035 (compared to 29% and 33% without IRA)	Increase of carbon-free electricity to 75–77% by 2030, ca. 90% by 2035 (compared to 75% in business as usual scenario)	**Gaps to net-zero pathway:** Growth in transmissions capacity 1.5–1.8 vs. 2.4%). Cumulative investment gap ($2.4–2.8tn vs. $3.2 tn)
Independent (Bistline et al.)	33–40% by 2030, 43–48% by 2035	Electricity contributes 38–80% of reductions 38–92% reduction in unabated coal generation by 2030. 32–52% ZEV sales by 2030. 11–32% decrease in fossil fuel use	Ambitions gap to US 2030 and 2050 climate targets; Uncertainty about external factors (growth, inflation). Non-cost factors (supply chains, permitting, network effects)

ZEV = zero emission vehicle.
Sources: DoE 2023, Jenkins et al. 2022, 2023 (REPEAT project), King et al. 2023 (Rhodium Group), Bistline et al. 2023.

the IRA's programs, many of its programs are indeed uncapped and could expand the actual amount of investment into much larger figures. Similar degrees of uncertainty are, obviously, associated with expected impacts of the IRA on job creation and positive effects for living standards and spending of individual households (summarized in Table 4.5 below).

4.4 Discussion: the disruptive politics of US green industrial policy

Many of the observations in the above case study point to the disruptive dynamics and effects of green industrial policy launched under the BIL and IRA, regarding not only the departure of the present from previous agendas and the imprint of political conflict on evolving policy images, but also the institutional framework and policy-making effects of current US climate policies. In order to summarize these observations, and based on our theoretical framework and the indicators proposed to evaluate vectors of policy development, insights from the present case study can be summarized in the following points:

- The *policy image* of climate action proposed by the Biden administration and subsequently enacted through the BIL and IRA has evolved in a context of political change, volatility, and contestation. Concerning its *scope*, the ambition to tackle the climate crisis as a cross-government challenge expressed in initial Executive Orders creates a rupture with the previous administration and establishes one of the key innovations of current policy-making in comparison even to failed previous attempts to establish a comprehensive approach to climate policy in the US. An intriguing aspect, however, is that the *priority* of climate targets in relation to other policy goals has been rendered ambiguous through the amalgamation of decarbonization goals with targets of economic prosperity, social justice, affordable healthcare and greater geopolitical independence. The *transformative appeal* of the policy images promoted by the BIL and IRA, therefore, is clear and ambitious but presents the reduction of GHG emissions primarily as a side effect of economic innovation and investment in social justice and security. As a whole, the sequence of policy images identified above is inseparable from the intense political conflict surrounding the negotiation

Table 4.5 Expected impacts of the Inflation Reduction Act in a wider policy-making context

	Evaluation of ongoing / future expenditure	Estimated macroeconomic effects (growth, jobs, energy security)	Effects on living and health standards / social factors
DoE	$369bn investment as reference value for IRA; combined expenditure in energy system of $430bn by IRA + BIL	44–59% decrease of net crude oil imports by 2030; push for deployment of new technologies (CCS, hydrogen, batteries)	$27–38bn in savings for spending on household energy bills (13–15% for companies); $500–1000 annual savings for individual households
DoT	Evaluation through Rhodium / MIT Clean Investment Monitor: concentration in disadvantaged areas Communication through invest.gov website	Creation of up to 1.5 million jobs in clean energy sectors	+80% of IRA investment in counties below national income average Fuel cost reduction for private vehicle owners by about 60%
REPEAT project	Cumulative capital investment $2.4–2.8tn 2023–35	Creation of 2.2–2.9mn new energy supply-related jobs by 2035 (breakdown by sectors) Doubling of annual rate of PV additions to 44-51GW 2023–30	Lowering annual energy expenditures 3–7%, $59–113bn for households Avoiding 18–37k premature deaths from exposure to fine particulate matter

PV = photovoltaic energy.
Sources: DoE 2023, Jenkins et al. 2022, 2023 (REPEAT project), King et al. 2023 (Rhodium Group).

of the legislative acts adopted as a consequence of the initial, more ambitious BBB agenda.

• As a consequence, the *policy venues* to be considered relevant for climate action in the US have evolved through a process of rapid change that has created an entirely new governance framework beyond its previous boundaries. *Vertical intervention* and guidance of the White House, together with the establishment of new positions and coordination mechanisms, is the decisive factor for the setup and operation of this framework that moves US climate action beyond the competence of the EPA. New executive departments of considerable political weight, particularly the DoT and DoE, are established as new key pillars of climate and energy policy-making. The *horizontal friction* observed between the White House and both chambers of Congress, particularly with regard to final negotiations about the IRA in the US Senate, are more intense than in the EU case but primarily concern the stage of policy formulation; since its adoption, the policy-making process launched through the IRA has been shielded relatively successfully from potential challenges, particularly through protracted budget negotiations and the intense polarization surrounding the US response to the Russian attack on Ukraine. The *public accountability* particularly of the IRA, however, is elevated through its perception as a signature achievement of the Biden administration and intense political contestation by key agents of the Republican party.

• In terms of *policy feedback*, the impact particularly of the IRA is open-ended but very likely to prompt considerable change in a wider range of policy-making areas and disruptive effects even beyond the borders of the US. The *speed* of decarbonization is set at a higher ambition than before and is likely to increase as an effect of adopted legislation; however, its precise trajectory is neither certain nor clearly prescribed beyond the legally nonbinding expression of goals in Executive Orders and the Nationally Determined Contribution submitted by the US after its re-entry into the framework of the Paris Agreement. The *stringency* of policy-making has assumed a new quality through the substantial addition of funds and instruments, while previously existing regulations concerning emission and efficiency standards have been upheld and reinforced rather than repealed. To reiterate, the absence of a regulatory framework prescribing reductions of carbon emissions and the open-ended, competitive dynamic of investment targeted by the IRA create a degree of uncertainty about its expectable effects, as demonstrated by the range of possible outcomes with regard to the expected expense, technological and energy mix and plausible effects on decarbonization. The effects of the current green industrial policy are very likely to outlast the current administration even in the case of a change in the presidency due to the longer-term effects of investment; however, modification or even repeal appears possible, qualifying the current US climate governance regime as less durable and more easily *reversible* than one established on the foundation of a climate law legally prescribing the achievement of carbon neutrality. Longer-term policy effects of the current legislation will, therefore, certainly be affected to a substantial degree by dynamics of competition, possible external shocks and political conflict.

To summarize this analysis, the launch, enactment and anticipated effects of the IRA in terms of policy change can be qualified as a source of disruptive policy and institutional change described as positive feedback by PET. An intriguing difference between policy-making dynamics, however, emerges between the process of enacting the IRA and its anticipated policy-making effects in later stages. As the previous analysis has shown, the establishment of the climate governance framework created through the IRA can serve as a text-book example of a policy breakthrough promoted by macropolitical agents and resulting in the involvement of new policy subsystems and decision-making processes; however, the envisaged longer-term effect of the IRA is based on the idea of incremental and self-reinforcing policy change driven by positive economic and political returns from investment in technologies and infrastructure that affect the trajectory of future paths of energy and economic policy-making.

This aspect is behind the references by President Biden to ribbon-cutting for new plants and industries as a factor that will create new paths of economic development and prove resilient against partisan polarization. What appears key for the future impact of US green industrial policy is therefore the transition from the still ongoing phase of roll-out of funds and policy initiatives from the broad policy package of the IRA and its subsequent materialization as an established factor of economic and energy policy-making at both the federal and US state levels. Considering the dynamics of punctuation identified in this case study, however, it is clear that many of the observations made here about the evolution of the US energy and climate governance framework create a notable contrast to the previous EU case study.

4.5 Conclusion: the ambiguous politics of climate action through green industrial policy

Through its focus on the IRA, this chapter has reviewed the historically most significant act of policy-making against climate change in the US. Throughout its analysis, the case study presented in this chapter has identified aspects of the policy-making process associated with disruption and political conflict. More specifically, it demonstrates asymmetries between two dimensions of the climate governance process prompted by the IRA.

First, in a policy-making dimension, the IRA is significant as a meaningful step towards more stringent action to decarbonize particularly the energy grid and mobility sector, and embedded in a longer-term ambition to achieve the net-zero target by mid-century. The size of available funds and scope of the institutional framework underline the disruptive impact and quality of the new policy-making framework that has been created to enact these ambitions. The launch of substantial investment in green technology and infrastructure is, however, not grounded in a regulatory framework establishing a mechanism for carbon pricing or legally binding commitments to reach prescribed GHG reduction targets. As a result of political constraints prohibiting the establishment of such a framework at the federal level, the success of the IRA as a breakthrough for action against climate change therefore has to rely on a problematic mix of market processes, the successful investment of public funds in targeted industries, and openly protectionist elements that remain open to external challenges and pushback.

Second, concerning the politics of climate change, the case study demonstrates how the linkage between political conflict around climate as a political issue, and its effect on subsequent policy-making, is far from straightforward. In this context, one of the insights from the analysis of the IRA is that its success has depended to a large degree on the amalgamation of climate targets with goals of economic growth, job creation and geopolitical independence. Broadening the space of climate action – understood as the expansion of its political and institutional framework from its original association with environmental policy to a broad governance process centered on industrial and economic policy – has fundamentally changed its perception and contributed to the diffusion and depoliticization of climate change as an issue rather than to increasing its visibility as a political priority.

The implications of this amalgamation of climate and related policy targets are twofold. On a political level, and building on our previous observations about the diffuse and ambiguous policy image of the IRA, extant poll data suggest that the bill has failed to be recognized as a significant step of action against global warming by the broad US public. In this regard, surveys such as the Yale Climate Opinion Map confirm that majorities of US citizens consider climate change as a problem; majorities also expect the federal government and president to do more against global warming and pursue the development of clean energy as a priority, even if opinions are split between supporters of both major parties particularly concerning the prioritization of climate change as a problem. At the same time, however, polls conducted on perceptions of the IRA suggest that close to half or even a clear majority of persons surveyed admitted to having heard nothing at all about its content, with only around 40% stating they had heard "a lot" or "some" about it (ClimateWire 2024b,i). In addition, another poll by AP reports that only about a quarter of respondents in the US state that the IRA's promotion of clean energy and vehicles had benefited them personally (ClimateWire 2024h). The same poll suggests that only around a third of US adults think that components of the IRA are helpful to address the problem of climate change (ibid.). This mismatch between the general recognition of climate change as a problem and the failure to successfully communicate the significance of extant policies to address its causes evidently creates a problem for the Biden administration to present itself as a champion of action against global warming.

References

Atkinson, Hugh. 2018. *The Politics of Climate Change Under President Obama*. London; New York: Routledge.

Bailey, Christopher J. 2015. *US Climate Change Policy*. Farnham, Surrey Burlington, VT: Ashgate.

Bang, Guri. 2021. "The United States: Conditions for Accelerating Decarbonisation in a Politically Divided Country." *International Environmental Agreements: Politics, Law and Economics* 21(1): 43–58. doi:10.1007/s10784-021-09530-x.

Barbanell, Melissa. 2022. "A Brief Summary of the Climate and Energy Provisions of the Inflation Reduction Act of 2022." www.wri.org/update/ brief-summary-climate-and-energy-provisions-inflation-reduction-act-2022 (June 12, 2024).

Bistline, John, Jeffrey Blanford, and Maxwell Brown. 2023. "Emissions and Energy Impacts of the Inflation Reduction Act Economy-Wide Emissions Drop 43 to 48% Below 2005 Levels by 2035 with Accelerated Clean Energy Deployment." *Science* 380(6625): 1324–27. doi:10.1126/science.adg3781.

Brewer, Thomas L. 2015. *The United States in a Warming World: The Political Economy of Government, Business, and Public Responses to Climate Change.* Cambridge: Cambridge University Press.

Carlson, Ann, and Dallas Burtraw, eds. 2019. *Lessons from the Clean Air Act. Building Durability and Adaptability into U.S. Climate and Energy Policy.* Cambridge: Cambridge University Press.

ClimateWire. 2022. "Biden Issues Executive Order Ahead of White House Climate Party." September 13.

ClimateWire. 2024a. "3 Things to Know About the Methane Fee." January 17, Vol. 10, No. 9. https://subscriber.politicopro.com/article/eenews/2024/ 01/17/3-things-to-know-about-the-methane-fee-00135877.

ClimateWire. 2024b. "Biden vs. Trump: Do Young Climate Voters Care?" February 16, Vol. 10, No. 9. www.politico.com/news/2024/02/18/joe-biden-climate-activists-00142037.

ClimateWire. 2024c. "Biden's Trick to Skirt Podesta Confirmation Fight." February 2. www.eenews.net/articles/bidens-trick-to-skirt-podesta-confi rmation-fight/.

ClimateWire. 2024d. "EPA Readies Reporting Rule for Methane Fee." April 19, Vol. 10, No. 9. www.eenews.net/articles/epa-readies-reporting-rule-for-methane-fee/.

ClimateWire. 2024e. "Harris, Regan to Roll out $20B EPA Climate Finance Program." April 3, Vol. 10, No. 9. www.eenews.net/articles/harris-regan-to-roll-out-20b-epa-climate-finance-program/.

ClimateWire. 2024f. "John Kerry Heads for the Exit." March 4, Vol. 10, No. 9. www.eenews.net/articles/john-kerry-heads-for-the-exit/.

ClimateWire. 2024g. "Podesta's First Day as US Global Climate Boss." March 7, Vol. 10, No. 9. www.eenews.net/articles/podestas-first-day-as-us-global-climate-boss/.

ClimateWire. 2024h. "Poll: Climate Change Concerns Grow, but Few Think Biden Law Will Help." April 19, Vol. 10, No. 9. https://apnews.com/article/ america-ap-norc-poll-climate-change-election-f9943ab6a37674ef2345f 4041618cf5e.

ClimateWire. 2024i. "Poll: Cutting Energy Prices More Powerful Campaign Message than Climate." February 16, Vol. 10, No. 9. doi:www.eenews. net/articles/poll-cutting-energy-prices-more-powerful-campaign-message-than-climate/.

ClimateWire. 2024j. "Strong Renewable Energy Growth Falls Short of Global Goal." January 11, Vol. 10, No. 9. https://subscriber.politicopro.com/arti cle/eenews/2024/01/11/strong-renewable-energy-growth-falls-short-of-glo bal-goal-00134796.

ClimateWire. 2024k. "US and Global Climate Policy: Can Podesta Do Both?" February 1, Vol. 10, No. 9. www.eenews.net/articles/us-and-global-clim ate-policy-can-podesta-do-both/.

ClimateWire. 2024l. "U.S. Emissions Fell 2% in 2023, Far Short of Climate Goals." January 10, Vol. 10, No. 9. https://subscriber.politicopro.com/arti cle/eenews/2024/01/10/u-s-emissions-fell-2-in-2023-far-short-of-climate-goals-00134524.

ClimateWire. 2024m. "US Starts 'Green Bank' to Finance Community Climate Projects." April 4, Vol. 10, No. 9. www.eenews.net/articles/us-sta rts-green-bank-to-finance-community-climate-projects/.

Collomb, Jean-Daniel. 2014. "The Ideology of Climate Change Denial in the United States." *European Journal of American Studies* 9(9–1). doi:10.4000/ejas.10305.

CRS (Congressional Research Service). 2022a. "Inflation Reduction Act of 2022 (IRA): Provisions Related to Climate Change." https://crsrepo rts.congress.gov/product/pdf/R/R47262#:~:text=IRA%20contains%20ei ght%20titles%2C%20each,resilience%20to%20climate%20change%20 impacts.

CRS (Congressional Research Service). 2022b. "Tax Provisions in the Inflation Reduction Act of 2022 (H.R. 5376)." https://crsreports.congress.gov/product/pdf/R/R47202.

Danish, Kyle. 2018. "Current Developments: North America." *Carbon & Climate Law Review* 1: 62–64.

Department of Energy. 2023. "Investing in American Energy. Significant Impacts of the Inflation Reduction Act and Bipartisan Infrastructure Law on the U.S. Energy Economy and Emissions Reductions." www.energy.gov/sites/default/files/2023-08/DOE%20OP%20Economy%20Wide%20R eport_0.pdf.

Dunlap, Riley E. 2019. "Partisan Polarization on the Environment Grows Under Trump." *Gallup.com.* https://news.gallup.com/opinion/gallup/248294/partisan-polarization-environment-grows-trump.aspx   (October 27, 2021).

Dunlap, Riley E., Aaron M. McCright, and Jerrod H. Yarosh. 2016. "The Political Divide on Climate Change: Partisan Polarization Widens in the U.S." *Environment: Science and Policy for Sustainable Development* 58(5): 4–23. doi:10.1080/00139157.2016.1208995.

EPA (Environmental Protection Agency). 2022a. "Year One Anniversary Report. Bipartisan Infrastructure Law." www.epa.gov/system/files/docume nts/2022-11/V-4_BIL_FirstAnniversaryReport_Nov142022.pdf.

EPA (Environmental Protection Agency), OAR. 2022b. "Summary of Inflation Reduction Act Provisions Related to Renewable Energy." www.epa.gov/green-power-markets/summary-inflation-reduction-act-provisions-related-renewable-energy (June 12, 2024).

Evergreen. 2022. "Evergreen Explains: The Climate Impact of the Inflation Reduction Act." www.evergreenaction.com/documents/The-Climate-Impact-of-the-IRA.pdf.

Federal Register. 2021a. "Executive Order 14008 of January 27, 2021: Tackling the Climate Crisis at Home and Abroad." www.govinfo.gov/content/pkg/FR-2021-02-01/pdf/2021-02177.pdf.

Federal Register. 2021b. "Executive Order 14052 of November 15, 2021: Implementation of the Infrastructure Investment and Jobs Act." www.govinfo.gov/content/pkg/FR-2021-11-18/pdf/2021-25286.pdf.

Federal Register. 2022a. "Executive Order 14072 of April 22, 2022: Strengthening the Nation's Forests, Communities, and Local Economies." www.govinfo.gov/content/pkg/FR-2022-04-27/pdf/2022-09138.pdf.

Federal Register. 2022b. "Executive Order 14082 of September 12, 2022: Implementation of the Energy and Infrastructure Provisions of the Inflation Reduction Act of 2022." www.govinfo.gov/content/pkg/FR-2022-09-16/pdf/2022-20210.pdf.

Federal Register. 2023. "Executive Order 14096 of April 21, 2023: Revitalizing Our Nation's Commitment to Environmental Justice for All." www.govinfo.gov/content/pkg/FR-2023-04-26/pdf/2023-08955.pdf.

Fiorino, Daniel J. 2022. "Climate Change and Right-Wing Populism in the United States." *Environmental Politics* 31(5): 801–19. doi:10.1080/09644016.2021.2018854.

Fisher, Dana R., Philip Leifeld, and Yoko Iwaki. 2013. "Mapping the Ideological Networks of American Climate Politics." *Climatic Change* 116(3): 523–45. doi:10.1007/s10584-012-0512-7.

Freeman, Jody. 2011. "The Obama Administration's National Auto Policy: Lessons from the 'Car Deal.'" *Harvard Environmental Law Review* 343(35).

Freeman, Jody. 2013. "Climate and Energy Policy in the Obama Administration." *Pace Environmental Law Review* 30(1): 375.

Glicksman, Robert. 2017. "The Fate of the Clean Power Plan in the Trump Era." *Carbon & Climate Law Review* 11(4): 292–302.

Gompertz, Dido. 2022. "Inflation Reduction Act Brings US in Reach of Climate Goals." *E3G*. www.e3g.org/news/inflation-reduction-act-brings-us-in-reach-of-climate-goals/ (June 12, 2024).

Gruenig, Max. 2023. "One Year Inflation Reduction Act. Initial Outcomes and Impacts for EU–US Trade and Investment." *E3G*. www.e3g.org/publications/one-year-inflation-reduction-act/ (June 12, 2024).

Guber, Deborah Lynn, Jeremiah Bohr, and Riley E. Dunlap. 2021. "'Time to Wake Up': Climate Change Advocacy in a Polarized Congress, 1996–2015." *Environmental Politics* 30(4): 538–58. doi:10.1080/09644016.2020.1786333.

Guliyev, Farid. 2020. "Trump's 'America First' Energy Policy, Contingency and the Reconfiguration of the Global Energy Order." *Energy Policy* 140: 1–10.

Jenkins, Jesse, Erin Mayfield, Jamil Farbes, Greg Schivley, Neha Patankar, and Ryan Jones. 2023. "Climate Progress and the 117th Congress: The Impacts of the Inflation Reduction Act and Infrastructure Investment and Jobs Act." doi:10.5281/zenodo.8087805.

Jenkins, Jesse, Erin Mayfield, Jamil Farbes, Ryan Jones, Neha Patankar, Qingyu Xu, and Greg Schivley. 2022. "Preliminary Report: The Climate and Energy Impacts of the Inflation Reduction Act of 2022." https://repeat project.org/docs/REPEAT_IRA_Prelminary_Report_2022-08-04.pdf.

Jiang, Betty. 2022. "US Inflation Reduction Act: A Tipping Point in Climate Action." https://scholar.google.com/scholar_lookup?title=%E2%80%9CUS+Inflation+Reduction+Act%3A+A+Tipping+Point+in+Climate+Action%E2%80%9D&author=B.+Jiang&publication_year=2022 (June 12, 2024).

Jotzo, Frank, Joanna Depledge, and Harald Winkler. 2018. "US and International Climate Policy Under President Trump." *Climate Policy* 18(7): 813–17.

Karapin, Roger. 2016. *Political Opportunities for Climate Policy: California, New York, and the Federal Government.* Cambridge: Cambridge University Press.

King, Ben, Hannah Kolus, Naveen Dasari, and Michael Gaffney. 2023. "Taking Stock 2023. US Emissions Projections After the Inflation Reduction Act." https://rhg.com/research/taking-stock-2023/.

King, Ben, John Larsen, and Hannah Kolus. 2022. "A Congressional Climate Breakthrough." https://rhg.com/research/inflation-reduction-act/.

Kramer, Ronald. 2020. "Rolling Back Climate Regulation: Trump's Assault on the Planet." *Journal of White Collar and Corporate Crime* 1(2): 123–30.

Larsen, John, Ben King, Hannah Kolus, Naveen Dasari, Galen Hiltbrand, and Whitney Herndon. 2022. "A Turning Point for US Climate Progress: Assessing the Climate and Clean Energy Provisions in the Inflation Reduction Act." https://rhg.com/research/climate-clean-energy-inflation-reduction-act/.

Leggett, Jane. 2019. "Potential Implications of U.S. Withdrawal from the Paris Agreement on Climate Change." https://sgp.fas.org/crs/misc/IF10668.pdf.

MacNeil, Robert. 2017. *Neoliberalism and Climate Policy in the United States. From Market Fetishism to the Developmental State.* Abingdon: Routledge.

Mann, Michael E. 2021. *The New Climate War: The Fight to Take Back Our Planet.* New York: Public Affairs, Hachette.

Marlon, Jennifer R., Emily Goddard, Peter D. Howe, Matto Mildenberger, Martial Jefferson, Eric Fine, and Anthony Leiserowitz. "Yale Climate Opinion Maps 2023." *Yale Program on Climate Change Communication.* https://climatecommunication.yale.edu/visualizations-data/ycom-us/ (June 12, 2024).

McCright, Aaron M., and Riley E. Dunlap. 2011. "The Politicization of Climate Change and Polarization in the American Public's Views of Global Warming, 2001–2010." *Sociological Quarterly* 52(2): 155–94. doi:10.1111/j.1533-8525.2011.01198.x.

Mehling, Michael. 2017. "A New Direction for US Climate Policy: Assessing the First 100 Days of Donald Trump's Presidency." *Carbon & Climate Law Review* 1: 3–23.

Mehling, Michael, and Antto Vihma. 2017. "'Mourning for America'. Donald Trump's Climate Change Policy." www.fiia.fi/wp-content/uploads/2017/10/analysis8_mourning_for_america-2.pdf.

Perls, Hannah. 2022. "Breaking Down the Environmental Justice Provisions in the 2022 Inflation Reduction Act – Harvard Law School." https://eelp.law.harvard.edu/2022/08/ira-ej-provisions/ (June 12, 2024).

Rabe, Barry. 2011. "Contested Federalism and American Climate Policy." *Publius: The Journal of Federalism* 41(3): 494–521.

Rabe, Barry. 2018. *Can We Price Carbon?* Cambridge, Massachusetts: The MIT Press.

Roy, Nicholas, Maya Domeshek, Dallas Burtraw, Karen Palmer, Kevin Rennert, Jhih-Shyang Shih, and Seth Villanueva. 2022. "Beyond Clean Energy: The Financial Incidence and Health Effects of the IRA." https://media.rff.org/documents/Report_22-11_v5.pdf.

Selby, Jan. 2019. "The Trump Presidency, Climate Change, and the Prospect of a Disorderly Energy Transition." *Review of International Studies* 45(3): 471–90.

Sussman, Glen, and Byron W. Daynes. 2013. *US Politics and Climate Change: Science Confronts Policy.* Boulder, CO [u.a.]: Lynne Rienner Publishers.

Thompson, Frank, Kenneth Wong, and Barry Rabe. 2020. *Trump, the Administrative Presidency, and Federalism.* Washington, DC: Brookings Institution Press.

UNFCCC. 2020. "The United States of America Nationally Determined Contribution Reducing Greenhouse Gases in the United States: A 2030 Emissions Target." https://unfccc.int/sites/default/files/NDC/2022-06/United%20States%20NDC%20April%2021%202021%20Final.pdf.

US Congress. 2021a. "Infrastructure Investment and Jobs Act – Summary." www.congress.gov/bill/117th-congress/house-bill/3684 (June 12, 2024).

US Congress. 2021b. "PUBLIC LAW 117–58—NOV. 15, 2021, 117th Congress ('Infrastructure Investment and Jobs Act')." www.congress.gov/117/plaws/publ58/PLAW-117publ58.pdf.

US Congress. 2022a. "Inflation Reduction Act of 2022 – Summary." www.congress.gov/bill/117th-congress/house-bill/5376 (June 12, 2024).

US Congress. 2022b. "PUBLIC LAW 117–169—AUG. 16, 2022, 117th Congress, ('Inflation Reduction Act')." www.congress.gov/117/plaws/publ169/PLAW-117publ169.pdf.

Vandeweerdt, Clara, Bart Kerremans, and Avery Cohn. 2016. "Climate Voting in the US Congress: The Power of Public Concern." *Environmental Politics* 25(2): 268–88. doi:10.1080/09644016.2016.1116651.

Vezirgiannidou, Sevasti-Eleni. 2013. "Climate and Energy Policy in the United States: The Battle of Ideas." *Environmental Politics* 22(4): 593–609. doi:10.1080/09644016.2013.806632.

White House. 2022. "Building a Better America. A Guidebook to the Bipartisan Infrastructure Law for State, Local, Tribal, and Territorial Governments, and Other Partners." www.whitehouse.gov/wp-content/uploads/2022/05/BUILDING-A-BETTER-AMERICA-V2.pdf.

White House. 2023. "Building a Clean Energy Economy: A Guidebook to the Inflation Reduction Act's Investments in Clean Energy and Climate Action." www.whitehouse.gov/wp-content/uploads/2022/12/Inflation-Reduction-Act-Guidebook.pdf.

White House. 2024. "The Biden–Harris Administration Immediate Priorities." www.whitehouse.gov/priorities/.

5 Conclusion

Climate agendas and instability in the European Union and United States

This volume presents one of the first comparative analyses of green recovery policies in the EU and US as the source of a major dynamic of change in policy-making on climate change: namely, its expansion to a cross-cutting agenda of economic and technological modernization that is enacted through a mix of public investment and creation of incentives for the deployment of zero-carbon infrastructure (Allan et al. 2021, Meckling 2021, Newell et al. 2021). The theoretical approach of the present analysis has focused on the interaction of two opposed vectors of change: namely, conditions to support a long-term and continuous progress of policies to achieve decarbonization described as policy stability, on the one hand; and sources of disruption caused by exogenous shocks and political contestation around climate change as a political issue, on the other.

From the theoretical perspective of PET, the recent development of climate governance in the two cases of the EU and US presents two contrasting dynamics: namely, a steady expansion and adaptation of proclaimed policy images, venues and instruments in response to a sequence of exogenous shocks that have resulted in incremental policy change on a prescribed path towards carbon neutrality in the EU, and a disruptive breakthrough to a new and large-scale approach to climate action evolving through a sequence of contested policy images and resulting in open-ended policy change not clearly prescribed by a regulatory framework for explicit targets of carbon reduction in the US. These findings resonate with and expand extant comparative perspectives on both cases (Skjaerseth 2013, Wurzel et al. 2021, Lewis 2021, Luterbacher & Sprinz 2018). What explains the occurrence of these different dynamics of policy change? Setting the present case

DOI: 10.4324/9781003452041-5

studies in their context of previous stages of policy development, key explanations proposed by PET can be applied to shed light on the present comparison of cases in the following respects.

First, our analysis has traced an evolution of policy images within a relatively short period of analysis: in the EU, as identified through a shift of climate agendas towards a more strategic and geopolitical rationale since the launch of the EGD agenda in 2019; and in the US, through the inclusion of zero-carbon targets into an agenda of economic modernization, recovery and social justice since the arrival in office of the Biden administration. The different dynamics driving these policy images in both cases are better clarified when set in longer-term perspective (Delbeke & Vis 2019, Brewer 2015, Kalantzakos 2017). The present study brackets a period of analysis that follows on almost two decades of policy initiatives to define rationales and targets of climate policy in the EU. In the case of the US, it is set in the aftermath of a highly polarized presidential election resulting in a transition between two administrations with highly visible and politically defining differences in their respective perspectives on climate change; it also continues a longer-term sequence of administrations with alternating views on climate change since the 1990s (Sussman & Daynes 2013). The more conditional and incremental adjustment of the EU climate agenda fits into this perspective of a longer and more continuous trajectory of climate policy-making in the EU as opposed to more volatile swings between contrasting policy approaches in the US.

Second, the present analysis corroborates assumptions by PET about the relevance of institutional linkages and interaction between agenda-setting at the macropolitical level and policy subsystems. In this regard, both case studies have identified shifts of attention between different political topics rather than steady attention to climate change at the level of macropolitical institutions; however, they also highlight different degrees of policy-making stability at the level of subsystems. In the EU case, the role of the Commission seems decisive as an agent that mediates pressures for policy change between macropolitical agenda-setting at the level of the European Council and the meso-level of its AWPs' policy-specific legislative proposals. The present analysis demonstrates the role of the Commission as a policy entrepreneur that is politically invested in the continuation of the EGD agenda and the role of policy beliefs advocated to promote the EGD agenda (Rietig 2019, Rietig & Dupont 2021). This role is enacted through its efforts for limiting the adaptation of policies to targeted revisions while upholding the pursuit of regulatory policy-making under the Fit for 55 program as a direct consequence of the adoption of the European Climate Law. By comparison, the effect of the investment agenda prompted by the BIL and IRA in the US case study has been enacted by direct hierarchical intervention through the central executive in the White House. This form of political direction has prompted a more disruptive shift to new and wider policy venues in comparison to the previous limitation of climate and energy policy to executive action centered on the EPA in a context of adversarial relations between states and the federal government (Thompson et al. 2020, Karapin 2016).

Finally, the present comparison of case studies confirms the link between observed forms of policy change and the permanence or breakup of established policy monopolies in subsystems with an established set of agents, decision-making procedures and fields of expertise. As the EU case study has demonstrated, its core subsystem in charge of regulatory legislation has remained in place as the key site for decision-making on the institutional and policy-related foundations of its EGD governance framework; this applied not only to the substantial regulatory components of the EU climate action agenda but also to the definition of principles and rules of governance mechanisms established for the management of recovery programs under the framework of the RRF. In line with the perspective adopted by PET, institutional change in the US links the policy changes brought about particularly by the IRA to the disruption of a previous policy monopoly centered on the EPA while re-directing it to the DoT, DoE and coordination mechanisms overseen by the White House.

To conclude this comparative discussion, we return to the broader question discussed at the outset: what insights can be drawn from the present analysis for the implications of policy stability versus punctuation for the progress of action against climate change (cp. Paterson et al. 2022)? At first sight, the comparison of the present two cases appears inconclusive in this respect: while the green recovery programs in both the EU and US have led to substantial policy progress towards the reduction of carbon emissions, their evolution is marked by different dynamics of stability and punctuation. In a general sense, this point suggests that no unequivocal link exists between degrees of political conflict and advances in climate policy. However, three more specific insights can be drawn from the present comparison of cases.

First, the present analysis demonstrates that arguments about the relevance of disruption and stability have to make a clearer distinction about two stages of the policy process: on the one hand, the stage of *climate policy creation* through the launch of new agendas and governance frameworks to pursue action against climate change, as primarily led by macropolitical agents; and on the other, the subsequent stage of *climate policy operation* involving the enactment of proposed measures through decision-making in specialized subsystems. Within this distinction, the present analysis suggests that the launch of a policy-making framework against climate change with no previous institutional foundation requires disruption; this point is demonstrated by developments in the US, where the most significant breakthrough for climate action at the federal level to date has been achieved only through protracted contentious negotiation against significant opposed interests. By contrast, the analysis of developments in the EU covers a case where the EGD agenda has been established as the continuation of a trajectory that reaches back to the conclusion of the first global agreements on climate change and puts a stronger emphasis on aspects of policy continuity.

Second, the reconstruction of developments in the US demonstrates the pitfalls and difficulties arising from an environment of intense political contestation for the promotion of climate policy. While the IRA is particularly significant as a breakthrough towards decarbonizing the energy grid and mobility, its passage comes with three risks that result from its stepwise evolution from a core component of the BBB agenda to a considerably reduced and rebranded piece of legislation: the difficulty of establishing it as a clearly targeted and understandable act of policy-making in the public perception given the multi-faceted and complex range of targets addressed by the legislation; its launch as a primarily incentive-driven policy with a relatively weak regulatory framework, primarily caused by its passage as a reconciliation bill to manage and overcome strong sources of opposition and conflict; and its relative vulnerability to blockades and repeal in future electoral cycles with potentially different political majorities in the White House and US Congress. This suggests that at the stage of policy operation, stability and positive feedback through political and economic returns from investment are more conducive for further progress towards decarbonization than protracted political conflict and disruption.

Finally, the theoretical framework applied here helps to distinguish between two related dimensions of the dualism of stability versus conflict: namely, aspects involved in its political dimension such as agenda-setting, macropolitical intervention in policy subsystems, and contestation between and within institutional arenas; and those relating to its policy-related dimension, particularly the continuity and irreversibility of policies enacted to realize climate targets. Further investigating the interaction of these two dimensions is a major task for research focusing on the shift of the political boundaries of climate governance, as initially addressed in analyses of Green Deal agendas in a context of various crises (Bloomfield & Steward 2020, Eckert 2021) and emerging research on green industrial policy programs (Allan et al. 2021, Lewis 2021, Meckling 2021, Newell et al. 2021).

Concerning limitations of the present analysis, the case studies have explored dynamics of change in the evolution of agenda-setting and policy images but left unaddressed the precise causes of change in comparison of the two cases. In this respect, pressures for change of climate agendas covered here cannot be related to the same sort of exogenous shock or source of political contestation: while challenges to the EGD agenda have primarily arisen from exogenous events such as the Covid pandemic, war in Ukraine and changing geopolitical environment in the case of the EU (Wendler 2023, Bongardt & Torres 2022), the turn towards a new climate-related agenda in the US is related more clearly to party political causes and the change of administration after the 2020 general election (Bang 2021, Jiang 2022, Kramer 2020). In its most familiar specification by Baumgartner and Jones, PET is relatively indifferent to such distinctions; these factors are considered with no particular distinction of whether spikes in the political salience of topics can be related to external events or developments within the political system such as changing majorities (Baumgartner et al. 2009: 39ff., 2018: 62ff.). The case comparison presented here suggests that party political factors are relevant for the rise of new policy images and subsequent dynamics of institutional and policy change beyond and independently from external shocks and resulting shifts in public attention.

Weighing implications for future research, the present study has implications that go beyond the specific analysis of climate policies in the EU and US. At a theoretical level, its main question is how to update accounts of climate policy integration (CPI) as a key concept for tracing and evaluating the linkages between climate and related policy targets (Mickwitz et al. 2009, Nilsson & Nilsson 2005, Tosun & Lang 2017). By referring to CPI, we address a concept that has received considerable attention in the literature about the evolution of EU climate governance: in the initial stages, as a concept closely related to environmental issues (Adelle & Russel 2013, Rietig 2013) but subsequently as a mechanism of major importance for evaluating EU climate and energy governance (Kettner & Kletzan-Slemanig 2020, Bocquillon & Maltby 2020, Dupont 2016) and the evolution of the EGD agenda (Oberthür & von Homeyer 2023, Holzleitner et al. 2019). The present study highlights two aspects of expansive policy-making dynamics not yet considered in large portions of this research literature: first, a clearer theory-based discussion of linkages and interactions between different stages of the policy process; and second, a more conflict- and process-driven understanding of mechanisms leading to the expansion of climate governance frameworks.

In a similar perspective, several recent accounts of CPI emphasize dynamic and agency-based aspects such as policy beliefs and political leadership as a factor for its success (Rietig 2019, Rietig & Dupont 2021). In this regard, the present study speaks to some of the most recent contributions to the literature on policy integration that move its focus from the evaluation of policy outcomes to aspects of political agency and conflict (Cejudo & Trein 2023a). This change in the understanding of the multi-faceted concept of policy integration (Tosun & Lang 2017) is associated with attempts to link concepts of policy integration more closely with theories of the policy process (Weible & Sabatier 2017, Weible & Workman 2022).

In this sense, one of the most widely received theoretical accounts of policy integration in recent years traces the integration of cross-cutting policy targets through four dimensions of the policy process: policy framing, subsystems interaction, policy goals and instruments (Candel & Biesbroek 2016, Biesbroek & Candel 2020). Through its perspective across these four dimensions, the approach highlights the often asynchronous, incoherent and conflictual progress of policy integration, including its failure and even reversal through dynamics of disintegration. Similarly, some recent contributions have addressed policy integration as a political process influenced by dynamics of politicization within the interaction of policy subsystems with macropolitical agents and decisions (Cejudo & Trein 2023a,b).

In the present study, the concept of CPI has not been introduced as a term of major relevance for the theoretical framework. However, it relates to this literature through its analysis of how dynamics of expansion of governance frameworks arise between stages of the policy process. Through this approach, the study can contribute to the research debate on policy integration in two respects: first, at a theoretical level, as a proposal for a set of detailed indicators to be applied for evaluating dynamics of change through the dimensions of agenda-setting, policy venues and subsequent decision-making results; and second, as an empirical demonstration of how dynamics of expansion unfold through these stages either through incremental or more disruptive dynamics. In this context, PET seems suited as a theory of the policy process that can explain sequences of disruptive change towards both policy integration and disintegration, especially in a perspective focused on the interaction between macropolitics and subsystems.

Turning to an outlook on future events, the present analysis was completed in a turbulent political year that is almost certain to establish new sources of conflict and disruption concerning the issue of climate change. Both the EU and US hold elections that feature policies related to climate change as one of their major political issues. As this book goes to press, elections to the new EP will have taken place that are widely expected to result in a further surge of right-wing populist and nationalist parties broadly opposing the EGD agenda or even denying climate change, as documented in the literature on the politics of far-right and populist parties on climate change (Marquardt & Lederer 2022, Huber et al. 2021, Forchtner 2020, Lockwood 2018). Probably the more important question concerning the 2024 EP election, however, is to what degree the political consensus of centrist parties will hold that granted its majority support to the von der Leyen Commission and subsequently formed legislative majorities to enact the European Climate Law and related legislation, particularly under the Fit for 55 package. In this regard, the future stance of the European People's Party (EPP) especially deserves further attention, not only at the level of future legislative debates within the EP but also concerning the stance of members of the European Council associated with the EPP towards the composition and leadership of the next EU Commission. Signs of fraying support for components of the EGD agenda within the EPP have been noticeable for some time, such as in the debate on the EU Nature Restoration Law or the more recent weakening of ecological provisions in the reform of the Common Agricultural Policy in early 2024. These fissures in the political support for the climate action agenda associated with the EGD deserve close scrutiny; however, based on the argument of the present study, it should be expected that the overall target and broad policy trajectory towards carbon neutrality adopted in the EU under the EGD agenda will be upheld even with more adverse political majorities in the EP.

In comparison, the political stakes are higher and the possible impact of election results more far-reaching in the case of the US presidential election. To what degree a victory of Republican candidate Donald Trump would lead to a reversal of current US climate policies is largely a matter of speculation at the present stage. The deeply ideological polarization between the two major parties and their candidates on the general issue of climate change aside, it is not evident to what extent a change in the presidency would lead to a full-scale repeal of climate-related policies. Three aspects deserve consideration.

First, most components of the IRA prescribe a time span for their enactment that reaches beyond the forthcoming election and would therefore require an explicit decision of repeal passed by Congress. Considering that the US states that have attracted large investment based on subsidies provided through the IRA include swing or even clearly Republican-leaning states, it is questionable how far a full repeal of current programs would receive sufficient support at this level. By contrast, influential US states such as California are likely to continue and fortify current climate policies even under a new federal government (ClimateWire 2023, 2024b). It therefore seems more likely that under a Republican majority and presidency, necessary action for efficient implementation of climate-related aspects (particularly in relation to siting and permitting), as well as cross-cutting provisions associated with social and environmental aspects of the bill, would be weakened or reversed. Actions aimed at a deliberate slow-walking of implementation and adjustment of green and social conditionality therefore appear as a more likely immediate prospect than a full-scale repeal (ClimateWire 2024b).

Second, some of the actions adopted by the Biden administration concerning investment decisions, supply chains and restrictions on trade are unlikely to be reversed and might even be strengthened under a future Trump presidency. In this regard, the establishment of punitive tariffs on the import of Chinese electric vehicles, but also of technology relevant for the green transition such as solar panels or semiconductors, is more likely to be complemented with additional and more openly protectionist restrictions on trade affecting goods and services not associated with clean energy and technology. The full implications for future policy agendas on climate change are difficult to assess but are almost certain to result in more volatility and a further re-orientation of climate action towards geopolitical rationales and agendas.

Finally, the clearest reversal of current policies is to be expected at the level of declarations and actions that can be established through direct executive action without explicit authorization by Congress: in particular, the set of Executive Orders pledging the commitment of the US to decarbonization targets and associated policy goals as discussed in the previous case study, but also the framework of regulations for GHG emission reductions from the energy and mobility sector established by the EPA. In this regard, it has been noted that the last year of the current term of President Biden has brought forward an unusually high number of executive decrees and regulations; these include a new set of more stringent rules to curb emissions from the energy sector and prompt the phasing out of fossil fuel-based power plants by the start of the next decade. While these provisions are certain to be subjected to review and repeal in a potential second Trump presidency, experience with the previous one suggests that longer time lags and legal challenges are associated with initiatives to repeal and might result in late and only partial reversals. A narrower degree of uncertainty about the continuation of US climate action accompanies the prospect of a presidency of current Vice-President Kamala Harris, arguably depending on majorities within Congress after the next US general election.

Beyond these two high-stakes electoral contests and regardless of their outcome, an almost certain prospect of climate policy-making is a continuation or even reinforcement of tensions between the EU and US already caused by components of the IRA (Kleimann 2023a,b). Another trend that is very likely to continue is the orientation of climate policy-makers towards a more uncertain, volatile and conflictual geopolitical context. In this regard, competition for investment in critical green technologies between the EU and US has become more aggressive, particularly through provisions of the IRA; indirect additional effects of this competition emerge as a result from sanctions and exclusion clauses directed against the entry of Chinese products into the US market, prompting exposure of the EU to further pressure on prices, and policy responses such as the imposition of import tariffs on electric vehicles. Against this background, formats for dialogue between the two entities such as the EU–US Trade and Technology Council seem only partly effective in responding to the more protectionist tendencies of current green industrial policy agendas; responses by the EU through programs such as those proposed in the GDIP currently appear as the politically more important response. Considering these aspects, the relationship between the EU and US as partners of the global climate regime continues to be fragile.

From this point of view, it seems clear that climate agendas in both the EU and US will continue to develop in a volatile global environment and in settings of intense political contestation. Even in this context, perhaps the most intriguing observation made in this volume is that in spite of its salience as a polarizing topic, climate change has become a political issue that is by now almost exclusively raised and debated in conjunction with one or several extant policy issues – most prominently, those of energy affordability and security, technological competitiveness and innovation – but also in relation to investment in public infrastructure and social equality issues. These policy linkages, and the way in which they shape agendas and policy processes to promote decarbonization, are quite likely the most relevant component in the future politics of climate change.

References

Adelle, Camilla, and Duncan Russel. 2013. "Climate Policy Integration: A Case of Déjà Vu?" *Environmental Policy and Governance* 23(1): 1–12. doi:10.1002/eet.1601.

Allan, Bentley, Joanna I. Lewis, and Thomas Oatley. 2021. "Green Industrial Policy and the Global Transformation of Climate Politics." *Global Environmental Politics* 21(4): 1–19. doi:10.1162/glep_a_00640.

Bang, Guri. 2021. "The United States: Conditions for Accelerating Decarbonisation in a Politically Divided Country." *International Environmental Agreements: Politics, Law and Economics* 21(1): 43–58. doi:10.1007/s10784-021-09530-x.

Baumgartner, Frank, Bryan Jones, and Peter B. Mortensen. 2018. "Punctuated Equilibrium Theory: Explaining Stability and Change in Public Policymaking." In *Theories of the Policy Process*, eds. Christopher M. Weible and Paul A. Sabatier. New York, NY: Westview Press, 55–102.

Baumgartner, Frank, Christian Breunig, Green Pedersen, Bryan Jones, Peter B. Mortensen, Michiel Nuytemans, and Stefaan Walgrave. 2009. "Punctuated Equilibrium in Comparative Perspective." *American Journal of Political Science* 53(3): 603–20.

Biesbroek, Robbert, and Jeroen J. L. Candel. 2020. "Mechanisms for Policy (Dis)Integration: Explaining Food Policy and Climate Change Adaptation Policy in the Netherlands." *Policy Sciences* 53(1): 61–84.

Bloomfield, Jon, and Fred Steward. 2020. "The Politics of the Green New Deal." *The Political Quarterly* 91(4): 770–79. doi:10.1111/1467-923X.12917.

Bocquillon, Pierre, and Tomas Maltby. 2020. "EU Energy Policy Integration as Embedded Intergovernmentalism: The Case of Energy Union Governance." *Journal of European Integration* 42(1): 39–57. doi:10.1080/07036337.2019.1708339.

Bongardt, Annette, and Francisco Torres. 2022. "The European Green Deal: More than an Exit Strategy to the Pandemic Crisis, a Building Block of a Sustainable European Economic Model*." *JCMS: Journal of Common Market Studies* 60(1): 170–85. doi:10.1111/jcms.13264.

Brewer, Thomas L. 2015. *The United States in a Warming World: The Political Economy of Government, Business, and Public Responses to Climate Change*. Cambridge: Cambridge University Press.

Candel, Jeroen J. L., and Robbert Biesbroek. 2016. "Toward a Processual Understanding of Policy Integration." *Policy Sciences* 49(3): 211–31. doi:10.1007/s11077-016-9248-y.

Cejudo, Guillermo M., and Philipp Trein. 2023a. "Pathways to Policy Integration: A Subsystem Approach." *Policy Sciences* 56(1): 9–27. doi:10.1007/s11077-022-09483-1.

Cejudo, Guillermo M., and Philipp Trein. 2023b. "Policy Integration as a Political Process." *Policy Sciences* 56(1): 3–8. doi:10.1007/s11077-023-09494-6.

ClimateWire. 2023. " 'Trump Insurance': How States Could Fortify Climate Policy in 2024." December 22, Vol. 10, No. 9. https://subscriber.politico pro.com/article/eenews/2023/12/22/trump-insurance-why-2024-could-see-states-entrench-climate-policy-00132960.

ClimateWire. 2024a. "California Is Preparing to Defend Itself – and the Nation – Against Trump 2.0." March 26, Vol. 10, No. 9. www.eenews.net/articles/california-is-preparing-to-defend-itself-and-the-nation-against-trump-2-0/.

ClimateWire. 2024b. "Podesta on Trump Undoing IRA: 'Very, Very Difficult.'" April 17, Vol. 10, No. 9. www.eenews.net/articles/podesta-on-trump-undoing-ira-very-very-difficult/.

Delbeke, Jos, and Peter Vis. 2019. *Towards a Climate-Neutral Europe: Curbing the Trend*. London: Routledge.

Dupont, Claire. 2016. *Climate Policy Integration into EU Energy Policy*. Abingdon, OX [u.a.]: Routledge.

Eckert, Sandra. 2021. "The European Green Deal and the EU's Regulatory Power in Times of Crisis." *JCMS: Journal of Common Market Studies* 59(S1): 81–91. doi:10.1111/jcms.13241.

Forchtner, Bernhard, ed. 2020. *The Far Right and the Environment: Politics, Discourse and Communication*. Abingdon: Routledge.

Holzleitner, Christian, Philip Owen, and Yvon Slingenberg. 2019. "Mainstreaming Climate Change in EU Policies." In *Towards Carbon Neutrality. Curbing the Trend*, eds. Jos Delbeke and Peter Vis. Abingdon: Routledge, 180–99.

Huber, Robert A., Tomas Maltby, Kacper Szulecki, and Stefan Ćetković. 2021. "Is Populism a Challenge to European Energy and Climate Policy? Empirical Evidence Across Varieties of Populism." *Journal of European Public Policy* 28(7): 998–1017. doi:10.1080/13501763.2021.1918214.

Jiang, Betty. 2022. "US Inflation Reduction Act: A Tipping Point in Climate Action." https://scholar.google.com/scholar_lookup?title=%E2%80%9CUS+Inflation+Reduction+Act%3A+A+Tipping+Point+in+Climate+Action%E2%80%9D&author=B.+Jiang&publication_year=2022 (June 12, 2024).

Kalantzakos, Sophia. 2017. *The EU, US and China Tackling Climate Change: Policies and Alliance for the Anthropocene*. London; New York: Routledge.

Karapin, Roger. 2016. *Political Opportunities for Climate Policy: California, New York, and the Federal Government*. Cambridge: Cambridge University Press.

Kettner, Claudia, and Daniela Kletzan-Slamanig. 2020. "Is There Climate Policy Integration in European Union Energy Efficiency and Renewable Energy Policies? Yes, No, Maybe." *Environmental Policy and Governance* 30(3): 141–50. doi:10.1002/eet.1880.

Kleimann, David, Niclas Poitiers, André Sapir, Simone Tagliapietra, Nicolas Véron, Reinhilde Veugelers, and Jeromin Zettelmeyer. 2023a. "Green Tech Race? The US Inflation Reduction Act and the EU Net Zero Industry Act." *The World Economy* 46(12): 3420–34. doi:10.1111/twec.13469.

Kleimann, David, Niclas Poitiers, André Sapir, Simone Tagliapietra, Nicolas Véron, Reinhilde Veugelers, and Jeromin Zettelmeyer. 2023b. "How Europe Should Answer the US Inflation Reduction Act." Bruegel Policy Contribution. Research Report. www.econstor.eu/handle/10419/274198 (June 10, 2024).

Kramer, Ronald. 2020. "Rolling Back Climate Regulation: Trump's Assault on the Planet." *Journal of White Collar and Corporate Crime* 1(2): 123–30.

Lewis, Joanna I. 2021. "Green Industrial Policy After Paris: Renewable Energy Policy Measures and Climate Goals." *Global Environmental Politics* 21(4): 42–63. doi:10.1162/glep_a_00636.

Lockwood, Matthew. 2018. "Right-Wing Populism and the Climate Change Agenda: Exploring the Linkages." *Environmental Politics* 27(4): 712–32. doi:10.1080/09644016.2018.1458411.

Luterbacher, Urs, and Detlef F. Sprinz, eds. 2018. *Global Climate Policy: Actors, Concepts, and Enduring Challenges*. Cambridge, Massachusetts: The MIT Press.

Marquardt, Jens, and Markus Lederer. 2022. "Politicizing Climate Change in Times of Populism: An Introduction." *Environmental Politics* 31(5): 735–54. doi:10.1080/09644016.2022.2083478.

Meckling, Jonas. 2021. "Making Industrial Policy Work for Decarbonization." *Global Environmental Politics* 21(4): 134–47. doi:10.1162/glep_a_00624.

Mickwitz, Per, Francisco Aix, Silke Beck, David Carss, and Nils Ferrand. 2009. *Climate Policy Integration, Coherence and Governance*. Helsinki: PEER. https://pure.au.dk/ws/files/56076592/PEER_Report2.pdf.

Newell, Peter, Matthew Paterson, and Martin Craig. 2021. "The Politics of Green Transformations: An Introduction to the Special Section." *New Political Economy* 26(6): 903–6. doi:10.1080/13563467.2020.1810215.

Nilsson, Måns, and Lars J. Nilsson. 2005. "Towards Climate Policy Integration in the EU: Evolving Dilemmas and Opportunities." *Climate Policy* 5(3): 363–76. doi:10.1080/14693062.2005.9685563.

Oberthür, Sebastian, and Ingmar von Homeyer. 2023. "From Emissions Trading to the European Green Deal: The Evolution of the Climate Policy Mix and Climate Policy Integration in the EU." *Journal of European Public Policy* 30(3): 445–68. doi:10.1080/13501763.2022.2120528.

Paterson, Matthew, Paul Tobin, and Stacy D. VanDeveer. 2022. "Climate Governance Antagonisms: Policy Stability and Repoliticization." *Global Environmental Politics* (Preprint) 22(2): 1–11. doi:10.1162/glep_a_00647.

Rietig, Katharina. 2013. "Sustainable Climate Policy Integration in the European Union." *Environmental Policy and Governance* 23(5): 297–310. doi:10.1002/eet.1616.

Rietig, Katharina. 2019. "The Importance of Compatible Beliefs for Effective Climate Policy Integration." *Environmental Politics* 28(2): 228–47. doi:10.1080/09644016.2019.1549781.

Rietig, Katharina, and Claire Dupont. 2021. "Presidential Leadership Styles and Institutional Capacity for Climate Policy Integration in the European Commission." *Policy and Society* 40(1): 19–36. doi:10.1080/14494035.2021.1936913.

Skjaerseth, Jon Birger, Guri Bang, and Miranda A. Schreurs. 2013. "Explaining Growing Climate Policy Differences Between the European Union and the United States." *Global Environmental Politics* 13(4): 61–80. doi:10.1162/GLEP_a_00198.

Sussman, Glen, and Byron W. Daynes. 2013. *US Politics and Climate Change: Science Confronts Policy*. Boulder, CO [u.a.]: Lynne Rienner Publishers.

Thompson, Frank, Kenneth Wong, and Barry Rabe. 2020. *Trump, the Administrative Presidency, and Federalism*. Washington, DC: Brookings Institution Press.

Tosun, Jale, and Achim Lang. 2017. "Policy Integration: Mapping the Different Concepts." *Policy Studies* 38(6): 553–70. doi:10.1080/01442872.2017.1339239.

Weible, Christopher M., and Paul A. Sabatier, eds. 2017. *Theories of the Policy Process*. New York: Westview.

Weible, Christopher M., and Samuel Workman, eds. 2022. *Methods of the Policy Process*. Abingdon: Routledge.

Wendler, Frank. 2023. "The European Green Deal Agenda After the Attack on Ukraine: Exogenous Shock Meets Policy-Making Stability." *Politics and Governance* Special Issue: Governing the EU Polycrisis: Institutional Change After the Pandemic and the War in Ukraine 11(4). doi:10.17645/pag.v11i4.7343.

Wurzel, Rüdiger, Mikael Skou Andersen, and Paul Tobin, eds. 2021. *Climate Governance Across the Globe. Pioneers, Leaders and Followers*. Abingdon, OX [u.a.]: Routledge.

Appendix

Policy documents used for the quantitative content analysis in Chapters 3 and 4

(1) European Union

(1a) European Council Conclusions, 2019–24

European Council. (2019a). "European Council (Art. 50) Meeting (13 December 2019) – Conclusions." EUCO XT 20027/19. www.consilium. europa.eu/media/41796/13-euco-art50-conclusions-en.pdf.

European Council. (2019b). "European Council Meeting (12 December 2019) – Conclusions." EUCO 29/19. www.consilium.europa.eu/media/ 41768/12-euco-final-conclusions-en.pdf.

European Council. (2019c). "European Council Meeting (17 and 18 October 2019) – Conclusions." EUCO 23/19. www.consilium.europa.eu/media/ 41123/17-18-euco-final-conclusions-en.pdf.

European Council. (2020a). "European Council Meeting (10 and 11 December 2020) – Conclusions." EUCO 22/20. www.consilium.europa.eu/media/ 47296/1011-12-20-euco-conclusions-en.pdf.

European Council. (2020b). "European Council Meeting (15 and 16 October 2020) – Conclusions." EUCO 15/20. www.consilium.europa.eu/media/ 46341/1516-10-20-euco-conclusions-en.pdf.

European Council. (2020c). "Special Meeting of the European Council (1 and 2 October 2020) – Conclusions." EUCO 13/20. www.consilium.europa.eu/ media/45910/021020-euco-final-conclusions.pdf.

European Council. (2020d). "Special Meeting of the European Council (17, 18, 19, 20 and 21 July 2020) – Conclusions." EUCO 10/20. www.consil ium.europa.eu/media/45109/210720-euco-final-conclusions-en.pdf.

European Council. (2021a). "European Council Meeting (16 December 2021) – Conclusions." EUCO 22/21. www.consilium.europa.eu/media/ 53575/20211216-euco-conclusions-en.pdf.

European Council. (2021b). "European Council Meeting (21 and 22 October 2021) – Conclusions." EUCO 17/21. www.consilium.europa.eu/media/ 52622/20211022-euco-conclusions-en.pdf.

European Council. (2021c). "European Council Meeting (24 and 25 June 2021) – Conclusions." EUCO 07/21. www.consilium.europa.eu/media/50763/2425-06-21-euco-conclusions-en.pdf.

European Council. (2021d). "Special Meeting of the European Council (24 and 25 May 2021) – Conclusions." EUCO 05721. www.consilium.europa.eu/media/49791/2425-05-21-euco-conclusions-en.pdf.

European Council. (2022a). "European Council Meeting (15 December 2022) – Conclusions." EUCO 34/22. www.consilium.europa.eu/media/60872/2022-12-15-euco-conclusions-en.pdf.

European Council. (2022b). "European Council Meeting (20 and 21 October 2022) – Conclusions." EUCO 31/22. www.consilium.europa.eu/media/59728/2022-10-2021-euco-conclusions-en.pdf.

European Council. (2022c). "European Council Meeting (23 and 24 June 2022) – Conclusions." EUCO 24/22.

European Council. (2022d). "European Council Meeting (24 and 25 March 2022) – Conclusions." EUCO 01/22. https://data.consilium.europa.eu/doc/document/ST-1-2022-INIT/en/pdf.

European Council. (2022e). "Special Meeting of the European Council (24 February 2022) – Conclusions." EUCO 18/22. www.consilium.europa.eu/media/54495/st00018-en22.pdf.

European Council. (2022f). "Special Meeting of the European Council (30 and 31 May 2022) – Conclusions." EUCO 21/22. www.consilium.europa.eu/media/56562/2022-05-30-31-euco-conclusions.pdf.

European Council. (2023a). "European Council Meeting (23 March 2023) – Conclusions." EUCO 04/23. https://data.consilium.europa.eu/doc/document/ST-4-2023-INIT/en/pdf.

European Council. (2023b). "European Council Meeting (29 and 30 June 2023) – Conclusions." EUCO 07/23. https://data.consilium.europa.eu/doc/document/ST-7-2023-INIT/en/pdf.

European Council. (2023c). "Special Meeting of the European Council (9 February 2023) – Conclusions." EUCO 01/23. https://data.consilium.europa.eu/doc/document/ST-1-2023-INIT/en/pdf.

European Council. (2023d). "European Council Meeting (26 and 27 October 2023) – Conclusions." EUCO 14/23. https://data.consilium.europa.eu/doc/document/ST-14-2023-INIT/en/pdf.

European Council. (2023e). "European Council Meeting (14 and 15 December 2023) – Conclusions." EUCO 20/23. https://data.consilium.europa.eu/doc/document/ST-20-2023-INIT/en/pdf.

European Council. (2024). "Special Meeting of the European Council (1 February 2024) – Conclusions." EUCO 2/24. www.consilium.europa.eu/media/69874/20240201-special-euco-conclusions-en.pdf.

(1b) President-elect and State of the Union addresses, 2019–23

European Commission. (2019). "2019 Speech by President-Elect von der Leyen in the European Parliament Plenary on the Occasion of the Presentation of Her College of Commissioners and Their Programme." DG Communication. https://ec.europa.eu/commission/presscorner/detail/en/speech_19_6408.

European Commission. (2020). "2020 State of the Union Address by President von der Leyen: Building the World We Want to Live in: A Union of Vitality in a World of Fragility." DG Communication. https://ec.europa.eu/commission/presscorner/detail/ov/SPEECH_20_1655.

European Commission. (2021). "2021 State of the Union Address by President von der Leyen: Strengthening the Soul of Our Union." DG Communication. https://ec.europa.eu/commission/presscorner/detail/ov/SPEECH_21_4701.

European Commission. (2022). "2022 State of the Union Address by President von der Leyen: A Union That Stands Strong Together." DG Communication. https://ec.europa.eu/commission/presscorner/detail/ov/SPEECH_22_5493.

European Commission. (2023). "2023 State of the Union Address by President von der Leyen: Answering the Call of History." https://ec.europa.eu/commission/presscorner/detail/ov/speech_23_4426.

(1c) European Commission Annual Work Programs

European Commission. (2020a). "Commission Work Programme 2020: A Union That Strives for More." COM (2020) 37 final, Brussels, 29.1.2020. https://eur-lex.europa.eu/resource.html?uri=cellar:7ae642ea-4340-11ea-b81b-01aa75ed71a1.0002.02/DOC_1&format=PDF.

European Commission. (2020b). "Commission Work Programme 2021: A Union of Vitality in a World of Fragility." COM (2020) 690 final, Brussels, 19.10.2020. https://eur-lex.europa.eu/resource.html?uri=cellar:91ce5c0f-12b6-11eb-9a54-01aa75ed71a1.0001.02/DOC_1&format=PDF.

European Commission. (2021). "Commission Work Programme 2022: Making Europe Stronger Together." COM (2021) 645 final, Strasbourg, 19.10.2021. https://commission.europa.eu/system/files/2023-01/cwp2022_en.pdf.

European Commission. (2022). "Commission Work Programme 2023: A Union Standing Firm and United." COM (2022) 548 final, Strasbourg, 18.10.2022. https://commission.europa.eu/system/files/2022-10/cwp_2023.pdf.

(2) United States

(2a) Inauguration and State of the Union addresses

White House. (2021). "Inaugural Address by President Joseph R. Biden, Jr."
WH Briefing Room. www.whitehouse.gov/briefing-room/speeches-rema
rks/2021/01/20/inaugural-address-by-president-joseph-r-biden-jr/.
White House. (2022). "Remarks of President Joe Biden – State of the Union
Address as Prepared for Delivery." WH Briefing Room. www.whitehouse.
gov/briefing-room/speeches-remarks/2022/03/01/remarks-of-president-
joe-biden-state-of-the-union-address-as-delivered/.
White House. (2023). "Remarks of President Joe Biden – State of the Union
Address as Prepared for Delivery." WH Briefing Room. www.whitehouse.
gov/briefing-room/speeches-remarks/2023/02/07/remarks-of-president-
joe-biden-state-of-the-union-address-as-prepared-for-delivery/.
White House. (2024). "Remarks of President Joe Biden – State of the Union
Address as Prepared for Delivery." WH Briefing Room. www.whitehouse.
gov/briefing-room/speeches-remarks/2024/03/07/remarks-of-president-
joe-biden-state-of-the-union-address-as-prepared-for-delivery-2/.

(2b) Spoken statements by the White House on BIL, IRA and climate change

White House. (2022a). "Remarks by President Biden on the Inflation
Reduction Act of 2022." WH Briefing Room. www.whitehouse.gov/brief
ing-room/speeches-remarks/2022/07/28/remarks-by-president-biden-on-
the-inflation-reduction-act-of-2022/.
White House. (2022b). "Remarks by Vice President Harris on Climate
Resilience." WH Briefing Room. www.whitehouse.gov/briefing-room/
speeches-remarks/2022/08/01/remarks-by-vice-president-harris-on-clim
ate-resilience/.
White House. (2022c). "Remarks by President Biden at Signing of H.R.
5376, The Inflation Reduction Act of 2022." WH Briefing Room. www.
whitehouse.gov/briefing-room/speeches-remarks/2022/08/16/remarks-
by-president-biden-at-signing-of-h-r-5376-the-inflation-reduction-act-of-
2022/#:~:text=We're%20cutting.   %20deficit%20to,on%20%2440%20
billion%20in%20profit.
White House. (2022d). "Remarks by President Biden on the Passage of H.R.
5376, The Inflation Reduction Act of 2022." WH Briefing Room. www.
whitehouse.gov/briefing-room/speeches-remarks/2022/09/13/remarks-by-
president-biden-on-the-passage-of-h-r-5376-the-inflation-reduction-act-of-
2022/#:~:text=This%20bill%20will%20lower%20the,(Applause.).

White House. (2022e). "Remarks by Vice President Harris at an Inflation Reduction Act Climate Event." WH Briefing Room. www.whitehouse.gov/briefing-room/speeches-remarks/2022/09/14/remarks-by-vice-president-harris-at-an-inflation-reduction-act-climate-event/.

White House. (2022f). "Remarks by Vice President Harris on the Historic Achievement of the Passage of H.R. 5376, The Inflation Reduction Act of 2022." WH Briefing Room. www.whitehouse.gov/briefing-room/speeches-remarks/2022/09/15/remarks-by-vice-president-harris-on-the-historic-achievement-of-the-passage-of-h-r-5376-the-inflation-reduction-act-of-2022/.

White House. (2023a). "Remarks by Vice President Harris in a Moderated Conversation on Climate." WH Briefing Room. www.whitehouse.gov/briefing-room/speeches-remarks/2023/03/09/remarks-by-vice-president-harris-in-a-moderated-conversation-on-climate-4/.

White House. (2023b). "Remarks by Vice President Harris on Combatting Climate Change and Building a Clean Energy Economy." WH Briefing Room. www.whitehouse.gov/briefing-room/speeches-remarks/2023/07/14/remarks-by-vice-president-harris-on-combatting-climate-change-and-building-a-clean-energy-economy/.

White House. (2023c). "Remarks by Vice President Harris at an Inflation Reduction Act Anniversary Event." WH Briefing Room. www.whitehouse.gov/briefing-room/speeches-remarks/2023/08/15/remarks-by-vice-president-harris-at-an-inflation-reduction-act-anniversary-event/#:~:text=And%20it%20is%20with%20these,this%20state%20knows%20so%20well.

White House. (2023d). "Remarks by President Biden on the Anniversary of the Inflation Reduction Act." WH Briefing Room. www.whitehouse.gov/briefing-room/speeches-remarks/2023/08/16/remarks-by-president-biden-on-the-anniversary-of-the-inflation-reduction-act/#:~:text=The%20law%20is%20going%20to,in%20impacts %20on%20climate%20change.

(2c) Written statements by the White House on BIL, IRA and climate change

White House. (2021). "President Biden Announces the Build Back Better Framework." WH Briefing Room. www.whitehouse.gov/briefing-room/statements-releases/2021/10/28/president-biden-announces-the-build-back-better-framework/.

White House. (2022a). "Statement from President Biden on Inflation Reduction Act of 2022." WH Briefing Room. www.whitehouse.gov/briefing-room/statements-releases/2022/07/27/statement-from-president-biden-on-inflation-reduction-act-of-2022/.

White House. (2022b). "Statement from President Biden on the Inflation Reduction Act." www.whitehouse.gov/briefing-room/statements-relea ses/2022/08/04/statement-from-president-biden-on-the-inflation-reduct ion-act/.

White House. (2022c). "Statement from Vice President Harris on the Inflation Reduction Act." WH Briefing Room. www.whitehouse.gov/briefing-room/ statements-releases/2022/08/12/statement-from-vice-president-harris-on-the-inflation-reduction-act/.

White House. (2022d). "Statement of Dr. Alondra Nelson on the Inflation Reduction Act Becoming Law." WH Briefing Room. www.whitehouse. gov/ostp/news-updates/2022/08/16/statement-of-dr-alondra-nelson-on-the-inflation-reduction-act-becoming-law/.

White House. (2022e). "President Biden Announces Senior Clean Energy and Climate Team." WH Briefing Room. www.whitehouse.gov/briefing-room/ statements-releases/2022/09/02/president-biden-announces-senior-clean-energy-and-climate-team/.

White House. (2022f). "Statement by NSC Spokesperson Adrienne Watson on Launch of the US–EU Task Force on the Inflation Reduction Act." WH Briefing Room. www.whitehouse.gov/briefing-room/statements-releases/ 2022/10/25/statement-by-nsc-spokesperson-adrienne-watson-on-launch-of-the-us-eu-task-force-on-the-inflation-reduction-act/.

White House. (2022g). "Biden-Harris Administration Releases Inflation Reduction Act Guidebook for Clean Energy and Climate Programs." WH Briefing Room. www.whitehouse.gov/briefing-room/statements-releases/ 2022/12/15/biden-harris-administration-releases-inflation-reduction-act-guidebook-for-clean-energy-and-climate-programs/.

White House. (2023a). "Biden-Harris Administration Releases Inflation Reduction Act Guidebook for Tribes." WH Briefing Room. www.whiteho use.gov/briefing-room/statements-releases/2023/04/04/biden-harris-adm inistration-releases-inflation-reduction-act-guidebook-for-tribes/.

White House. (2023b). "Treasury Releases New Guidance, Strengthening Incentives for Domestic Clean Energy Manufacturing." WH Briefing Room. www.whitehouse.gov/cleanenergy/clean-energy-updates/2023/05/ 12/treasury-releases-new-guidance-strengthening-incentives-for-domestic-clean-energy-manufacturing/.

(2d) Fact sheets by the White House on IRA, BIL and climate change

White House. (2022a). "FACT SHEET: President Biden Signs Executive Order to Strengthen America's Forests, Boost Wildfire Resilience, and Combat Global Deforestation." WH Briefing Room. www.whitehouse.gov/ briefing-room/statements-releases/2022/04/22/fact-sheet-president-biden-signs-executive-order-to-strengthen-americas-forests-boost-wildfire-resilie nce-and-combat-global-deforestation/.

White House. (2022b). "FACT SHEET: 10 Ways the Biden-Harris Administration Is Making America Resilient to Climate Change." WH Briefing Room. www.whitehouse.gov/briefing-room/statements-releases/2022/08/01/fact-sheet-10-ways-the-biden-harris-administration-is-making-america-resilient-to-climate-change/.

White House. (2022c). "BY THE NUMBERS: The Inflation Reduction Act." WH Briefing Room. www.whitehouse.gov/briefing-room/statements-releases/2022/08/15/by-the-numbers-the-inflation-reduction-act/.

White House. (2022d). "FACT SHEET: How the Inflation Reduction Act Helps Black Communities." WH Briefing Room. www.whitehouse.gov/briefing-room/statements-releases/2022/08/16/fact-sheet-how-the-inflation-reduction-act-helps-black-communities/.

White House. (2022e). "FACT SHEET: How the Inflation Reduction Act Helps Latino Communities." WH Briefing Room. www.whitehouse.gov/briefing-room/statements-releases/2022/08/16/fact-sheet-how-the-inflation-reduction-act-helps-latino-communities/.

White House. (2022f). "FACT SHEET: How the Inflation Reduction Act Helps Rural Communities." WH Briefing Room. www.whitehouse.gov/briefing-room/statements-releases/2022/08/17/fact-sheet-how-the-inflation-reduction-act-helps-rural-communities/.

White House. (2022g). "FACT SHEET: How the Inflation Reduction Act Helps Tribal Communities." WH Briefing Room. www.whitehouse.gov/briefing-room/statements-releases/2022/08/18/fact-sheet-how-the-inflation-reduction-act-helps-tribal-communities/#:~:text=TRIBAL%2DSPECIFIC%20FUNDING,clean%20energy%20production%20and%20use.

White House. (2023). "FACT SHEET: One Year In, President Biden's Inflation Reduction Act is Driving Historic Climate Action and Investing in America to Create Good Paying Jobs and Reduce Costs." WH Briefing Room. www.whitehouse.gov/briefing-room/statements-releases/2023/08/16/fact-sheet-one-year-in-president-bidens-inflation-reduction-act-is-driving-historic-climate-action-and-investing-in-america-to-create-good-paying-jobs-and-reduce-costs/.

(2e) Policy guidance documents on BIL and IRA issued by the White House

White House. (2022). "Building a Better America. A Guidebook to the Bipartisan Infrastructure Law for State, Local, Tribal, and Territorial Governments, and Other Partners." WH Office. www.whitehouse.gov/wp-content/uploads/2022/05/BUILDING-A-BETTER-AMERICA-V2.pdf.

White House. (2023). "Building a Clean Energy Economy: A Guidebook to the Inflation Reduction Act's Investment in Clean Energy and Climate Action; Version 2." WH Office on Clean Energy Innovation and Implementation. www.whitehouse.gov/wp-content/uploads/2022/12/Inflation-Reduction-Act-Guidebook.pdf.

(2f) Text of BIL and IRA legislation

US Congress. (2021). "PUBLIC LAW 117–58—NOV. 15, 2021, 117th Congress ("Infrastructure Investment and Jobs Act")." www.congress.gov/117/plaws/publ58/PLAW-117publ58.pdf.

US Congress. (2022). "PUBLIC LAW 117–169—AUG. 16, 2022, 117th Congress, ("Inflation Reduction Act")." www.congress.gov/117/plaws/publ169/PLAW-117publ169.pdf.

Keywords used for quantitative content analysis

(1) European Union

Thematic category	Climate and environment	Economy and industry	Recovery and social issues	Security and external relations	Energy
Keywords used for dictionary	Climate change	Industry	RRF	Ukraine	Renewable energy
	Forest	Competitiveness	Covid-19	Russia, Russian	Fossil fuel
	Environmental	Technology	Just transition	High representative	Energy efficiency
	Green deal	Circular economy	Pandemic	Security and defence	Clean energy
	Biodiversity	Raw material	Recovery and resilience	Aggression	Energy transition
	Climate neutrality	State aid	Cohesion policy	European defence	Energy system
	Greenhouse gas	Green transition	Next generation	International cooperation	Energy price
	Environment	Research	Gender equality	Strategic autonomy	Energy innovation
	Climate-neutral	Renewable hydrogen	Social right	External action	Energy source
	Global warm	Industrial strategy	Fundamental right	International partner	Energy union
	CO_2 emission	Digital transition	Energy poverty	Partnership with	Energy security
	Climate law	Artificial intelligence			Energy market
	CBAM / carbon border adjustment mechanism	Private investment			Natural gas
		Capital market			
		Digital age			
		Digital transformation			

(2) United States

Thematic category	Climate and environment	Economy and jobs	Social security and justice	Health	Clean energy
Keywords used for dictionary	Climate crisis	Energy security	Environmental justice	Prescription	Solar
	Climate change	Economy	Social security	Health insurance	Heat pump
	Climate resilience	Job	Disadvantaged community	Mental health	Wind turbine
	Extreme weather	Wage	Special interest	Drug	Energy economy
	Extreme heat	Infrastructure	Underserved community	Health	Energy efficient
	Climate change	Efficiency	Low-income community	Medicare	Energy efficiency
	Water	Electric vehicle	Justice40	Insulin	Clean energy
	Forest	Technology	Community	Cancer	
	Emission	Investment	Tribal	Uninsured	
	Pollution	Worker	Family	Healthcare	
	Wildfire	Corporation	Child	Medicaid	
	Greenhouse	Manufacture	Good-paying		
	Environmental	Business	Uninsured		
	Flood	Economic	Wage		
	Drought	Company	Education		
			Equity		

Index